# THE TIMELY AND THE TIMELESS

THE JOHN DEWEY SOCIETY LECTURE SERIES

The John Dewey Lecture is delivered annually under the sponsorship of the John Dewey Society at the annual meeting of the National Society of College Teachers of Education. This book is an elaboration of the Lecture given on that occasion. The intention of the Series is to provide a setting where able thinkers from various sectors of our intellectual life can direct their most searching thought to problems that involve the relation of education to culture. Arrangements for the presentation and publication of the Lecture are under the direction of the John Dewey Society Commission on Lectures.

*The John Dewey Society Lecture—Number Eleven*

# THE TIMELY AND THE TIMELESS:

## *The Interrelationships of Science, Education, and Society*

*BY*

Bentley Glass

ACADEMIC VICE PRESIDENT AND DISTINGUISHED PROFESSOR OF BIOLOGY, STATE UNIVERSITY OF NEW YORK AT STONY BROOK, STONY BROOK, NEW YORK

*FOREWORD BY*

Ward Madden

CHAIRMAN, COMMISSION ON LECTURES, THE JOHN DEWEY SOCIETY FOR THE STUDY OF EDUCATION AND CULTURE

*Basic Books, Inc., Publishers*

NEW YORK LONDON

Library of Congress Catalog Card Number: 78–116849
SBN 465–08536–9
Manufactured in the United States of America

# FOREWORD

BY WARD MADDEN
*Chairman, Commission on Lectures*
*The John Dewey Society for the Study of Education and Culture*

SCIENCE IS EXUBERANTLY FLOODING the world with knowledge. As Bentley Glass points out, we can expect hundred-fold, if not thousand-fold, increases in scientific productivity in the course of a single generation. Science is transforming both our minds and our daily existence. It is revolutionizing our concepts of the universe, life, man, and man's destiny. At the same time it is producing a thoroughgoing technological reordering of our habits and activities.

And yet, strangely, there is a countercurrent to the scientific tide—a current now running swiftly and deeply through society. The tide-rip whirlpools of social and personal restlessness are seen everywhere. It is not simply a direct confrontation of science and antiscience—although this too is visible. Rather, much more deeply submerged in society's depths, there is an increasing tension between the established culture, of which science is only one part, and a "counter culture," to use the term made prominent by Theodore Roszak, whose ideas are themselves a manifestation of the culture he describes.

Science is now a part of the established culture, and the established culture seeks to make rational both the conduct and the understanding of life. Reason, reasonableness, logical order, organization, and analysis are imperative ideals of the established culture, which has been thousands of years in the making. Thus Aristotle's apotheosis of man's rationality. Thus the development of theology, by the Scholastics and their predecessors, as a logical and analytic ordering and explaining of the religious experience, which in the Far East, by contrast, remained essentially unanalyzable—a direct involvement of the whole, feeling person. Thus the development of parliamentary forms of government, in the attempt to replace the arbitrary irrationalities of despotic rule with reasonable debate and with decision on the basis of open access to relevant fact. Thus the growth of jurisprudence, with its substitution of rules of evidence for magical signs of guilt. Thus the rationalization of production within the factory by the division of labor, the devising of the assembly line, the Taylorization of management practices, and finally computerized automation. Thus the large-scale corporate organization of the economy, which, regardless of the ideological form which "rationalizes" it, makes rational planning on a national scale possible at least to a degree.

And thus the rise of science, which, whatever else it is, is an effort to achieve a rational understanding of the experienced world. If the counter culture has placed the established culture on the defensive, science too, as part of the established culture, is on the defensive, even in the midst of its success.

The rebellions against the established culture take many forms, seemingly different, yet mysteriously interrelated. For example, there is a psychedelic quality in the black revolution, there is soul in campus radicalism, and there is an Esalenizing of personal encounter in the breakdown of the distinction between actors and audience in contemporary theater.

The counter culture is suspicious of the life of reason. What counts is the fullness of the present moment, intensity of feeling, honesty and wholeheartedness of response to the other person, uncorrupted by cool planning or calculated concern for advantage.

And the established culture is vulnerable. It has trouble in distinguishing between the reasoned life and the rationalized life. The mechanistic quality of industrial order and efficiency has a numbing and stupefying effect upon the human psyche; law and order become rationalizations for oppression; the objective knowledge of science is made to seem more valid than the personal knowledge expressed in poetry.

Yet the counter culture is no more discriminating. In stressing the primacy of feeling over reason, of aesthetic values over utility, of personal authenticity over the public presentation of a pseudo-self, of interpersonal tenderness, care, and concern over social atomism, the counter culture does locate weak points in the established culture. The man from underground is finding the brittle fragilities in the crystal palace and knocking a shattered maze of cracks into it by doing little more than thumb his nose. Yet, in insisting on the priority of feeling *over* reason, of spontaneity *over* discipline, of freedom *over* order, he wrecks his own ideals. By indiscriminately throwing out the life of reason along with illegitimate rationality, he gives himself a license to live carelessly, dogmatically, self-destructively, and often ruthlessly.

Although not necessarily the main target, science gets into the line of fire by seeming to the counter cultural revolutionary to be irrelevant to and uncaring of human concerns, and, even on its own grounds, to be distorting reality by presenting a too objectivist picture of it. The psychedelic visionary's dreams are for him as empirically real as are the conceptual constructs of science—in fact more so, because the latter *are* constructs and as such are at least once removed from direct experience. This is not to say that the counter culture is necessarily drug-addicted—the example is simply symbolic of its epistemological trend.

The trouble, perhaps, is that science's success has been only superficial. It has never really become a part of the shared mentality of men, not even in advanced countries. Most men have been consumers of the products of science but have acquired very little understanding of the spirit, methods, or basic concepts of science. This is the condition that concerns Bentley Glass, and he is at pains in this book to show both the need for and ways of attaining a greater penetration of society, not by the artifacts of science, but by its basic approach and ideas. He is sensitive to the fact that neither the established culture nor the counter culture is sufficiently aware of the distinction between a life of reason and illegitimate rationality. He believes that science, properly understood by both its practitioners and by people generally, can contribute to the living of a good life.

Bentley Glass writes in the tradition of the great scientific humanists of the past, but with the advantage of standing upon their shoulders and addressing himself to the peculiar difficulties of the contemporary scene. He sees education at all levels from nursery through graduate school as having a fundamental role to play in helping men learn to approach their problems scientifically yet humanely and sensitively. But it must be a radically redesigned education, stimulated by first-rate scientific minds and by innovating science education centers such as the ones he has observed in Japan. It must be an education which stresses the spirit of discovery and methods of inquiry. These give both science and life their vitality, interest, and productiveness. Furthermore, large organizing concepts and unifying themes, not troops of regimented facts and definitions, must characterize the content of science education. Only when thus armed with adequate methodological equipment and relevant, viable concepts will the ordinary man be ready to play an effective role in helping the entire society become more reasonable and humane.

But, more than that, education must help science itself transcend its own limitations—unassimilable mountains of new knowledge, overspecialization, a too limited ethical and social sensitivity. He predicts, in effect, that if these limitations are not overcome, there will be a massive turning away from and attack upon science by the counter culture.

But one can believe, with Bentley Glass, that if men, with the help of education, can learn to apply the methods and concepts of science, properly conceived, to the whole range of human problems, there is hope that the established culture and the counter culture will combine their best characteristics, will slough off their worst, and will become, along with the scientific enterprise, part of a new culture truly devoted to the life of reason *and* feeling.

# CONTENTS

# THE TIMELY AND THE TIMELESS

# 1

# THE TIMELY AND THE TIMELESS: The Content of Education

### *Science and Education*

Perhaps two million years ago, at the mouth of his cave shelter in Olduvai Canyon in East Africa, a gnarled and weatherbeaten old man of a species ancestral to our own sat in the sun. In one hand he held a large nodule of flint, in the other a smaller hammer-stone with a point. Beside him crouched a boy, gazing intently at the old man's motions. A sharp blow of the hammer-stone on the side of the flint nodule detached a rough, flat flake that fell to the ground. The operation was repeated, over and over again, until a crude tool or weapon, pointed at one end and of a size to be held conveniently in the hand, had been produced. The old man then gave the boy a fresh nodule and the hammer-stone and grunted an encouragement to him to try also. Carefully the boy began, heeding the sounds of approval and disapprobation which the old man made as he watched, and thus the boy learned that certain kinds of rocks split and flake readily when struck in certain places, and that other kinds of rocks are better as hammer-stones. It must have been in this way that education first began, with the transmission of some skill from the older generation to the younger—doubtless before language had developed fully, in a species not yet quite man.

Much, much later, around the beginning of the last Ice Age some 75,000 years ago, a small company of Neanderthal men, women, and children gathered at Le Moustier, in what is now southern France, around a shallow grave they had excavated. Here, on a pillow made of a pile of flake tools and weapons, they carefully disposed the body of their companion, an eighteen-year-old youth who had met with some mishap or had died of disease. They placed the body on its side, resting as in sleep on his right arm, and gently covered him with earth. No doubt they mourned, yet what is most significant is that the burial in such a manner bespoke a belief in a life hereafter where the youth would have need of his weapons and his tools. When we think of the peoples of ancient but historical civilizations, among whom there was a universal practice of burying tools, clothing, and jewelry with the dead in firm belief in renewed life in some nether world, there can be little doubt that Neanderthal men had already begun to speculate about the why and wherefore of human existence and to weave about themselves the lore of gods and evil spirits to explain the mysterious and uncontrollable features of their world. The second great aspect of education, at first quite separate from the transmission of skills, had emerged. Yet was it truly quite so separate, even then? The empirical skills, the fruit of trial and error and the outcome of keen observation, augmented man's power over the prey he hunted and the enemy he fought. By Neanderthal times he dared to stand in combat even with the gigantic cave bear who competed for his dwelling place. But the lore of myth and magic, the rites of religion, and the speculations about the nature of the world and of man were likewise intended to enlarge human power, through binding in mystic rite and religious devotion the alliance of man with the supernatural which he saw everywhere about him.

It seems that in early civilizations the two streams of education were nevertheless regarded as quite distinct. The empirical tradition flowered in the arts and crafts of mankind, but it was often relegated to women and to slaves. The mystical stream flowing through religion and philosophy was for priests and rulers, and in its beginnings "liberal education" was the education of the free man in philosophy, religion, letters, rhetoric, and law, not a training in navigation or in manufactures, commerce, or agriculture which were the basis of everyday life. Both streams merged in the development of modern science. From trial and error the empirical stream became the arts and crafts, and these gave place to a scientifically based technology. The philosophical stream began in magic, superstition, and primitive religion, was enveloped in wars and persecutions, and flowered into the cosmology of Copernicus and Galileo, the laws of Newton, the evolutionary theory of Darwin, and the psychology of Pavlov and Freud. Today science dominates the practice of agriculture, medicine, and technology. In equal measure it is pre-eminent in shaping man's conceptions of himself and his place in nature, his hopes, his doubts, and his God. As Carl Becker has so eloquently said:

The cosmic view of the universe of infinite spaces, and of man's ultimate fate within it, is man's achievement—the farthest point yet reached in the progressive expansion of human intelligence and power. It is not rightly to be taken as a description of events that are relevant to man's purposes, but rather as an ideal result of those purposes—the manifestation of his insatiable curiosity, his indefeasible determination to know. As such it is less an objective world of fact than man's creation of the world in his own image. It is in truth man's most ingenious invention, his supreme work of art.[1]

The functions of science in society thus embrace and affect all education. On the one hand, science enlarges the freedom of the mind through liberation from the shackles of ignorance and superstition. On the other hand, it enlarges man's vision and expands his horizons by extending choice and opportunity. To paraphrase Lyman Bryson, man is not free to choose what he has never heard of or what doesn't exist.[2] Prior to the invention of printing, as I have written elsewhere,[3] few men could choose to read a book; and prior to the invention of television one could not choose between reading the book and watching a television program. Until now no man could visualize our blue and cloud-enswirled earth rising over the desolate horizon of the moon; yet recently, by means of spaceship and color television, millions have actually seen it in its entrancing beauty, a small globe of loveliness floating in the blackness of space.

[1] Carl Becker, *Progress and Power* (Stanford: Stanford University Press, 1936), p. 102. Caravelle Edition (New York: Vintage Books, Random House, 1965), p. 116.

[2] Lyman Bryson, *Science and Freedom* (New York: Columbia University Press, 1947), p. 5.

[3] Bentley Glass, *Science and Liberal Education* (Baton Rouge: Louisiana State University Press, 1959), p. 60.

The methods of science constitute the most powerful means yet devised by man for extending his limited senses and for distinguishing reality from hallucination. The mind of man, seen collectively, promotes through science an ever finer adaptation to the human environment, and when man himself proves insufficiently adaptable, it molds the environment itself into an agreeable artificiality. Science, through its technological applications, has clearly become the greatest force for change in the conditions of human existence and is accelerating that change almost beyond further endurance. Science has been among the greatest liberating, liberalizing forces in human thought. It is indispensable in the education of modern man, who so desperately seeks adjustment to his frenetic civilized environment.

Yet not only does it embody the seeds of change and of progress, it also engenders the most critical problems of modern society—the insanity of military power, the bursting human populations, the exploitation of unrenewable resources, the destruction of the environment by pollution of air, soil, and water, the control of behavior by drugs, and the lengthening human life span. The very limitations of science constitute formidable dangers posed by the unthinking scientism that would make of natural science a complete and sufficient view of human thought and action. Science tells us much, but never everything. It is restricted to matter and energy, space and form and time. It cannot weigh or measure values; it is thwarted by intangibles. Although it records the truths of experience, it never reveals the whole of truth. Its noblest, currently most satisfying concepts may tomorrow require gross correction or may fail altogether to explain new phenomena. Science ever strives for the objective, the confirmable, while the inner life of each of us remains, and always must remain, subjective, well-nigh incommunicable. Science is the product of the human mind, but what the mind is we cannot say, beyond identifying it with total behavior.

As Bertrand Russell well said, science can enhance among men two great evils, war and tyranny, since the powers of science can be used equally for good or evil ends.[4] What irony, that what may be, and indeed has been, quite the greatest force in liberating the human mind can also be turned to enslave man and destroy him! Yet that danger is posed by all power, not by that alone which grows out of the increasing understanding of man and nature. Man now stands with formerly unimaginable physical and biological powers in his grasp, and he is desperately afraid. It is of himself that he is afraid, of the choices he must make, of the unexplored consequences of those choices, of his own blindness and his own greed.

Education, which enables each generation to build on the achievements of the past and which nurtures the seeds of powers yet to be, must then reflect the significance of the natural sciences as the root of both material change and of widening human perception and understanding, whence derive its fundamental and supportive relationships to the economy, the social organization, and the political philosophy of each nation. It must also treat the natural sciences as the root of imminent disasters, of possible social collapse, of ultimate doom. I thus dare assert that history is no history without some revelation of the role of the sciences and of technology in the ascent of man to his present state of power. The social

[4] Bertrand Russell, *The Impact of Science on Society* (New York: Columbia University Press, 1951), p. 51.

sciences are no sciences without a grasp of the problems that grow out of the technological uses of scientific discovery. Thirty years ago, John Dewey said regarding the place of science in education:

> The full victory will not be won until every subject and every lesson is taught in connection with its bearing upon creation and growth of the kind of power of observation, inquiry, reflection and testing that are the heart of scientific intelligence. Experimental philosophy is at one with the genuine spirit of a scientific attitude in the endeavor to obtain for scientific method this central place in education.
>
> Finally, the sciences and philosophy of education can and should work together in overcoming the split between knowledge and action, between theory and practice, which now affects both education and society so seriously and harmfully.[5]

Science and education must each deal in harmony with the timely and the timeless. The former includes not only all empirical knowledge but more broadly everything that is relevant to the relation of the individual to society. In Dewey's epochal book *The School and Society* (1899), he asserted that the unity of all the sciences is found in "geography," which he would therefore make the focus of the curriculum. Reading his description of what he meant by geography, it is plain to see that it was intended in no restrictive sense, but rather expanded into all that we today would call *human ecology,* the relations of man to every aspect of the physical, biological, and social environment. What indeed could be more central, more timely than to examine as fully as possi-

[5] John Dewey, "The Relation of Science and Philosophy as the Basis of Education," *School and Society,* 47 (1938): 470–473. Reprinted in *John Dewey on Education* (New York: The Modern Library, 1964), pp. 15–19. See p. 19.

ble the sources and the limits of man's power as he molds nature to his will, recklessly building and destroying, purging and polluting, without heed for the long-range consequences?

But the timeless is no less important—that noble view of man's own place in the universe of which Becker spoke, the whole range of human values that have emerged from our animal natures through evolution and have been idealized and sublimated as man has become an increasingly social creature. With Pascal we can say, "Man has this superiority: He knows that the universe can with a breath destroy him, yet at the moment of death he knows that he dies, and knows also the advantage which the universe thereby has over him; but of all that the universe knows nothing." Through his discoveries in the last few decades of his million years of existence as a species, man has become the master of energies that can destroy not only himself and all of his works, but all life on earth, perhaps even the terrestrial globe itself. He has gained knowledge of evolutionary processes that can be used to create new species and, if he wishes, to remodel himself. He knows the most secret details of heredity, and perhaps is on the verge of creating life itself. Has he also the wisdom to attain a higher plane of values, to reconcile immediate benefit with the pursuit of long-range goals? Can he curb the greed of the living for the sake of the yet unborn, a thousand generations hence? Man is a creature of vast contradictions, selfish and loving, aggressive and cooperative, crass and idealistic, powerful and helpless. In one quality alone, it may be, his nature finds its unity: he is possessed, as Becker said, of insatiable curiosity, of an indefeasible determination to know. Out of this there burgeon both his science, compact of methods for finding and testing the truth about himself and his world, and his processes of education, whereby he communicates his findings to others in order that each generation may stand on the shoulders of all its predecessors.

What are the bounds and limits of understanding? In the university each student and each teacher, each scientist, scholar, and philosopher is devoted to the advancement of knowledge, not simply to its preservation. Education is thus bonded to research, and the fruits of research through technology become the service of the university to the society of which it is a part and upon which it must depend for support. The data and the facts alone do not constitute knowledge, in the sense of understanding. Information is needful, but the observations must be fitted into concepts and conceptual schemes, or paradigms, that determine one's outlook and direct one's processes of investigation and inquiry. Just as the study of a science must penetrate beyond its data and its laws and embrace also its methods of inquiry and its historical process, so too must education embrace the timely, the relevant, the social context, and also the timeless, the quest for the nature of things and especially the nature of the great enigma, the existence of man, the "I" who views the majesty of the universe. Of that total understanding there are as yet no certain margins, either at the extremity of the billion visible galaxies or at the bounds of the elementary particles. Here is still the "endless frontier" to challenge the curiosity of man, to enlarge the scope of education.

## *Content or Process—Subject Matter or Method of Inquiry*

Recent criticisms of the new science curricula in physics, chemistry, biology, and the earth sciences have revived old issues regarding the objectives of science education and the best mean of attaining them. The argument commonly takes the form of supposed alternatives: content or process; subject matter or method of inquiry. J. Myron Atkin has described this duality in the following words:

No conception of science is complete unless we are aware that one of these parts represents an attempt to comprehend and explain nothing less than the entire universe and what happens in it, by seeking to derive fundamental generalizations that account for motion, for life, for all the changes we know. And no conception of science is complete unless we are also aware that a second element represents this search for understanding as a human activity, an activity that engages men. . . .

The former element of science . . . is sometimes called the "content" of science, or, less commonly, the "products" of science, or the "major concepts" of science.

The latter element is sometimes called "process." But it has had other names over the decades such as "problem solving," "scientific method," and "inquiry." [6]

Clear as this distinction is, something is yet lacking if it is thought to characterize sufficiently the total educational role of science. The natural sciences have changed medicine, agriculture, and nutrition from a primitive reliance upon intuitive art, tradition, and empiricism to the status of applied sciences. The technologies of industry and commerce have likewise been rendered scientific, are increasingly based upon scientific advances. As Joseph J. Schwab has said:

Science has taken over the role of empire almost completely.[7]

Industrial democracy has made science the foundation of

[6] J. Myron Atkin, "A Critical Look at 'Process' in Science Education," *EPIE Forum,* 1, No. 8 (April 1968): 6–10. See p. 6.

[7] Joseph J. Schwab, "The Teaching of Science as Enquiry," in *The Teaching of Science,* by Joseph J. Schwab and Paul F. Brandwein (Cambridge: Harvard University Press, 1962), p. 19.

national power and productivity. Science now plays the part once played by exploration, by empire, and by colonial exploitation. It is the change in social role which has changed the pattern of scientific enquiry.[8]

Change has become the major feature of human civilization through an increasingly rapid development of enormous powers to modify and control raw nature; and the advance of science is the principal factor in this technological revolution. Education in science must therefore increasingly concern itself with other alternative goals: topical and social relevance or timeless natural law? The economic, political, and cultural impacts of science or man's place in the universe?

Already concerned with this problem by the turn of the present century, John Dewey spoke emphatically of the dangers of science teaching centered wholly on subject matter. In 1910 he stated that:

> . . . Science teaching has suffered because science has been so frequently presented just as so much ready-made knowledge, so much subject-matter of fact and law, rather than as the effective method of inquiry into any subject-matter.[9]

A few years later (1916), he returned to the subject of the relation of science to subject matter:

> Science . . . consists of the special appliances and methods which the race has slowly worked out in order to conduct reflection under conditions whereby its procedures and results are tested. It is artificial (an acquired art), not spontaneous; learned, not native. To this fact is due the unique, the invalu-

[8] *Ibid.*, p. 18.

[9] John Dewey, "Science as Subject-Matter and as Method," *Science*, 31 (1910): 121–127. Reprinted in *John Dewey on Education*, p. 187.

able place of science in education, and also the dangers which threaten its right use. Without initiation into the scientific spirit one is not in possession of the best tools which humanity has so far devised for effectively directed reflection. One in that case not merely conducts inquiry and learning without the use of the best instruments, but fails to understand the full meaning of knowledge. . . . On the other hand, the fact that science marks the perfecting of knowing in highly specialized conditions of technique renders its results, taken by themselves, remote from ordinary experience—a quality of aloofness that is popularly designated by the term abstract. . . .

Science has been defined in terms of method of inquiry and testing. At first sight, this definition may seem opposed to the current conception that science is organized or systematized knowledge. The opposition, however, is only seeming . . . [for] scientific subject matter is organized with specific reference to the successful conduct of the enterprise of discovery, to knowing as a specialized undertaking.[10]

Dewey could not have been more positive and emphatic about the matter than he was in the earlier of these two essays. He affirmed in ringing terms that:

One of the only two articles that remain in my creed of life is that the future of our civilization depends upon the widening spread and deepening hold of the scientific habit of mind; and that the problem of problems in our education is therefore to discover how to mature and make effective this scientific habit.[11]

And again,

. . . only the gradual replacing of a literary by a scientific education can assure to man the progressive amelioration of his lot. Unless we master things, we shall continue to be mastered

[10] John Dewey, "The Nature of Subject-Matter," in *Democracy and Education* (New York: The Macmillan Company, 1916), pp. 223–224. Reprinted in *John Dewey on Education,* pp. 369–370.

[11] John Dewey, "Science as Subject-Matter and as Method." Reprinted in *John Dewey on Education,* p. 191.

by them: . . . science, not words, casts the only compelling spell upon things.[12]

Was Dewey then advocating a study of process and method without content, without subject matter except for the process itself? His lectures on "The School and Society" in 1899 make it quite evident that that was not at all what he intended. In his paean to "geography" in the first of the three lectures entitled "The School and Social Progress," Dewey clearly indicated that what he sought was a principle for selecting content.

The unity of all the sciences is found in geography. The significance of geography is that it presents the earth as the enduring home of the occupations of man. The world without its relationship to human activity is less than a world.[13]

He goes on to characterize geography in terms that make it clear, as I have already said, that he had in mind what today might be more frequently termed "human ecology," the relation of man in every way to the complex environment, both inanimate and living, within which he dwells and which he bends to his own purposes. In science education the selection of content should illuminate these relations by firsthand experience that brings the child into contact with realities. Moreover, said Dewey:

Our social life has undergone a thorough and radical change. If our education is to have any meaning for life, it must pass through an equally complete transformation.[14]

[12] *Ibid.*

[13] John Dewey, *The School and Society* (Chicago: University of Chicago Press, 1899). Reprinted in Martin S. Dworkin (ed.), *Dewey on Education* (New York: Teachers College, Columbia University, 1959), p. 42.

[14] *Ibid.*, p. 49.

It is most significant to reflect that when John Dewey wrote those words the age of the automobile and the airplane had not yet begun. The command of nuclear energy, whether for weaponry or industrial power, was unimagined. Radio, in its infancy, was limited to the use of ships at sea, and radio broadcasting had not commenced. Television was undreamed of. No vitamin had been identified; no hormone had been discovered. Mendel's discoveries still lay forgotten. The development of hybrid corn and other achievements of plant breeding that have assured America so bountiful a supply of food were not yet seen in the realm of theory, let alone of actuality. There were no sulfa drugs or antibiotics to control infectious disease. The identification of viruses as widespread causes of human disease had scarcely begun. It was not known how hereditary changes, that is, mutations, occur, or that X rays or certain chemical agents could greatly multiply their frequency. The thought that man might one day be able to control the speed and direction of evolutionary processes was remote. Biochemistry was in its infancy. Little was yet known of protein structure, enzyme action, antibody production, or the significance of nucleic acids. Consequently there was little understanding of the constitutional basis of disease or congenital defect, of resistance to infections and toxins, or of the nature of the hereditary material of living things. The average length of life in the United States was about twenty-five years less than now. How immeasurably our vision of the nature of the universe and of the nature of life has been widened! How greatly the shape of human activity and the structure of society have changed in the intervening decades! How vastly, because of the achievements of science and technology, our perils and our problems have proliferated and increased in severity. If John Dewey in 1899 thought of man's social life as undergoing thorough and radical change, and of education as consequently in need of an equally complete transformation, how much truer the thought, how much greater the need, at the end of these seven ensuing decades!

In subsequent years few educational authorities seem to have struck a concordant note on this issue. There has been a general tendency to decry the teaching of science as a mere body of facts, as simply accumulated knowledge; but there is less agreement on the substitute. James Bryant Conant, in his book *On Understanding Science,* and in the foreword to the *Harvard Case Histories in Experimental Science,* has defined science as:

> . . . a series of concepts or conceptual schemes arising out of experiment or observation and leading to new experiments and observations.[15]

The facts of science, he added, grow out of the experiments and observations and are: ". . . tied together by the concepts and conceptual schemes of modern science." [16] To him the effective teaching of science is directed toward an understanding of the major concepts and conceptual schemes in each science. This is best achieved by a case study of the way in which the path of science meanders toward increased understanding through the contributions of many men who engage in observation and experiment, making errors and cor-

[15] James Bryant Conant, *On Understanding Science,* (New Haven: Yale University Press, 1947), pp. 24–25; also in Foreword: *Harvard Case Histories in Experimental Science* (Cambridge: Harvard University Press, 1950), p. 4.

[16] *Ibid.*

recting errors, employing speculative ideas and selecting facts to build working hypotheses, and testing their deductions from the conceptual schemes. The practical arts move from pure empiricism toward applied science, which reduces the amount of empiricism, until eventually the concepts and conceptual schemes of pure science are guiding the further advancement of the practical arts.

The emphasis on the importance in learning of the major conceptual schemes of a science is rephrased by Jerome S. Bruner, in *The Process of Education,* in terms of the *structure* of a subject. He has pointed out that:

> The scientists constructing curricula in physics and mathematics have been highly mindful of the problem of teaching the structure of their subjects, and it may be that their early successes have been due to this emphasis. Their emphasis upon structure has stimulated students of the learning process.[17]

A few years later (1964), the National Science Teachers Association brought together a group of scientists who formulated seven major conceptual schemes underlying all the sciences and coupled these with five major aspects of the process of science, which ought also to be kept in mind by writers and curriculum planners. I have elsewhere criticized the selection of these conceptual schemes as a basis for developing biological curricula,[18] for which they scarcely provide sufficient scope and specificity, but it must be said that for the physical sciences they might indeed serve very well. In any case, we find in this publication, *Theory into Action,* first in

[17] Jerome S. Bruner, *The Process of Education* (Cambridge: Harvard University Press, 1960), p. 8.

[18] Bentley Glass, "Theory into Action—a Critique," *The Science Teacher,* 32, No. 5 (1965): 20–30, 82–83.

Paul DeHart Hurd's prefatory essay on a theory of science education consistent with modern science, again in the coupling of the seven conceptual schemes with the five aspects of the process of science, enumerated and discussed by Ernest Nagel, and finally in the essay on the local implementation of science curriculum development, a recognition of the need to emphasize both the structure and conceptual schemes on the one hand and the processes of scientific investigation and inquiry on the other hand.[19]

As for the need to teach science as inquiry, no one has emphasized it more strongly than Joseph J. Schwab. The need, he has written, is:

> . . . ironically enough, that science be taught as science. What is required is that in the very near future a substantial segment of our publics become cognizant of science as a product of *fluid enquiry,* understand that it is a mode of investigation which rests on conceptual innovation, proceeds through uncertainty and failure, and eventuates in knowledge which is contingent, dubitable, and hard to come by. It is necessary that our publics become aware of the needs and conditions of such enquiry and inured to the anxieties and the disappointments which attend it.[20]

In many of the science curricula developed in the United States since 1957, both those for secondary and those for elementary school science, a strong effort has been made to acquaint the student with the processes of science by active involvement and participation in investigation and inquiry. Some critics, for example, J. Myron Atkin, believe that the

[19] *Theory into Action . . . in Science Curriculum Development* (Washington: National Science Teachers Association, 1964).

[20] Schwab, *The Teaching of Science,* pp. 4–5.

idea has in fact been overdone in certain curricula—in particular, in the elementary school program of the American Association for the Advancement of Science known as "Science—A Process Approach." Atkin criticizes the failure of this program to give proper attention to the basic concepts on which science rests, while instead it undertakes to improve general skills that can be developed in isolation from the context of any specific discipline, that are assumed to be transferable to the study of any other scientific discipline, and that follow naturally in a certain developmental sequence. Atkin's basic criticism is that the schematic processes of science, so taught, have very little to do with the way in which scientists in fact do work:

> A basic flaw in the process is the apparent assumption that science is a sort of commonsensical activity, and that the appropriate "skills" are the *primary* ingredients in doing productive work. There seems to be no explicit recognition of the powerful role of the conceptual frames of reference within which scientists and children operate and to which they are firmly bound. These general views of the physical world demand careful nurture and modification by a variety of means.[21]

Atkin therefore prefers the view that science education should introduce the child to new conceptual frames of reference, new ways of looking at the universe.

George Basalla, on the other hand, has recently attacked the new science curricula on the ground that they are concerned ultimately with content, not process. He does not refer to "Science—a Process Approach," but directs his fire at the high school science curricula. He says:

[21] Atkin, "A Critical Look at 'Process' in Science Education," p. 9.

The major failures of the new science curricula [are] failure to comprehend the nature of science and technology; failure to give serious consideration to the social effects of science and technology; and failure to include unambiguously the social sciences within the limits of scientific activity.[22]

It is quite evident that by "process" he has in mind two quite different aspects of the study of science, aspects which tend to become confused. Although Basalla starts out by talking about the failure of the new curricula to engage in a study of the nature of inquiry, he moves on to the nature of technology, its relation to science, and the impact of both on society. Now the nature of science as a social and historical process is surely not the same as the nature of science as an individual scientist's process of investigation. Both are indeed important. Can both be taught effectively at the same time?

## *The Educational Philosophy of a Modern Science Curriculum*

My own experience in these matters has over the past decade grown so largely out of my connection with the Biological Sciences Curriculum Study that I hope I may be pardoned for devoting particular attention to it as an example of the new spirit which has entered the teaching of science in our times. The BSCS is only one of quite a number of science curriculum programs supported by the National Science Foundation since 1957, but it has several features which make it particularly worthy of attention.

In the first place, because it is commonly taught in the

[22] George Basalla, "Science, Society, and Science Education," *Bulletin of the Atomic Scientists,* 24 (June 1968): 45–48.

tenth grade of our secondary schools, biology is the first science met by the student upon passing from the junior high school into the senior high school, and thus about 80 per cent of all high school students enroll in biology. Because only one year of science in senior high school is normally required for graduation or entrance into college, the percentage of students continuing into eleventh-grade chemistry falls to about 50 per cent, and the percentage enrolling in twelfth-grade physics is scarcely 20 per cent. For a large proportion of our general population, biology is consequently the only experience of science they meet at the upper secondary level.

In the second place, the four alternative programs prepared by the BSCS for tenth-grade biology have been very widely adopted by our public and private schools since 1963, when the first commercially available editions were published. Over one and a half million of these books—the so-called Blue, Green, and Yellow Versions, which present biology from different points of view and with different emphases, and the Patterns and Processes course, which is designed specifically for students with learning difficulties—have been sold and are currently in use. Since the total number of each cohort of biology students in the tenth grade is about two and a half million, perhaps over half of all the biology programs in the U.S. high school are now using the BSCS books. No other science curriculum is so widely used in the U.S. high schools.

In the third place, no American science curriculum project has drawn wider attention from abroad. Over fifty countries have prepared adaptations or are currently doing so, and in Latin America, Japan, Taiwan, the Philippines, Israel, and Australia outstanding BSCS programs are under way and new books are now available. The enthusiasm with which local groups of science teachers and biologists in these countries have adopted the essence of the BSCS program makes it especially significant to examine the nature of that program and the philosophy that has guided it.

Like the physics and chemistry curriculum studies (PSSC, CBA and Chem-Study), the creation of the BSCS followed a great concern aroused in the scientific community in the second half of the 1950's, first because of the dwindling supply, in proportion to the rapidly increasing need, of scientists and technologists in our society, and soon thereafter because of the orbiting of the first Sputnik, which awakened scientists, educators, and government officials to a realization that the United States was in grave danger of being surpassed in scientific and technological world leadership. Professor Jerrold Zacharias of the Massachusetts Institute of Technology vigorously advocated as the best course of action to counter such threats a thoroughgoing reform and revitalization of secondary school science teaching. For the first time in this century a mutual concern drew able and distinguished scientists out of their laboratories and classrooms to participate in the educational revolution, not simply as advisers and critics but as actual day laborers in an unfamiliar field.

The first imperative faced by the biologists of the BSCS was the necessity to cope with the antiquated content of the biology courses in the high schools. In large measure the failure to modernize the curriculum is because of the extraordinary increase in scientific knowledge in our century. Biology is a particularly rapidly developing area of science. In fact, it has become a mighty congeries of sciences which today no single biologist can pretend to encompass. By the end of this century the fund of biological knowledge will have redoubled seven, or even ten, times, to amount to definitely more than one hundred times, and perhaps even more than one thousand times, what it was in 1900. Entirely new biological sciences have arisen, and older ones have been modified beyond recognition.

Biology in the high schools was even more seriously antiquated than chemistry or physics. It had never received quite the administrative sanction of the physical sciences. To provide laboratories was not routine policy. Many biology classes met in ordinary classrooms, at best provided with tables on which experiments might have been performed had there been proper facilities—water, electricity, gas. Instead, it was only too obvious that biology was supposed to be limited to gross or microscopic observations or dissections of living or pickled materials. That the biological sciences in the past fifty years had moved away from the purely observational, descriptive aspects of science to the truly experimental was not recognized by the schools. The content of the textbooks and courses in the 1950's differed in no evident respect from what it had been in the 1920's. The principles and laws, the concepts and conceptual schemes which were included were those of a bygone, archaic day. Other great ideas had never been admitted at all. The great theme of organic evolution, which is central to the organization and interpretation of biology, and which has become vastly developed in the past forty years beyond the simple Darwinian schema, was disregarded and most often remained unmentioned. The fear that a book with the horrid word "evolution" in it could not be sold to school districts in large parts of the United States was quite sufficient to suppress it. Sometimes one found a euphemistic reference to the "theory of organic development"; more often evolution was simply excluded. The pressure of social and religious views was so effective that Hermann Joseph Muller was led to exclaim in indignation, "One hundred years without Darwin is enough!" To a nonbiologist this may seem a real, but perhaps not an overwhelming, defect. One must realize the centrality of evolution in the biological sciences to appreciate the consequences. One must understand that modern genetics makes no sense without it, the relation of organism to environment makes no sense without it, regulation and adaptation of form and function make no sense without it. The theory of evolution is not only the key to the understanding of the past of life on earth, it is the key to the understanding of the socially critical phenomenon of race, and it is the key to man's future. Within man's grasp there is now the power to mold his own species, as well as all others, according to design. This power grows from an understanding of genetic and evolutionary principles. It poses serious social and ethical problems. Are we once again to blunder into an age of frightful unforeseen powers without knowledge? Is it not still true that "without vision the people perish"?

Another biological matter on which the biology textbooks of the 1950's were silent was the subject of human sexual reproduction. Yet another was race. These were among the subjects banned as unspeakable. Many others, representing the noblest achievements made in our time in the understanding of life, were omitted simply because they were too new or too difficult to introduce in the context of the old and familiar. Molecular biology, the child of biochemistry and genetics, themselves children of the twentieth century, had been born and had matured. Environmental biology had become the science of populations and communities and lifted a cry of impending doom because of man's heedless forgetfulness that he is the creature of his environment and must in time pay the penalty for overcrowding it, defiling it, or destroying it. But both molecular biology and environmental biology were conspicuously absent in the high school courses.

Since the books of the 1920's were already of good size, the curriculum revisers of the 1960's could not simply add the significant new knowledge to the old books. Not even a careful integration and synthesis of the new with the old would serve, although integration and synthesis are necessary. Nothing short of a complete restudy and reorganization of biological knowledge would do. There would have to be a ruthless pruning of older information, including favorite examples and still valid concepts, as well as invalid principles and outworn conceptual schemes. The first task was boldly to say: "We can in fact present modern biology meaningfully and intelligibly to the high school student and we will make a fresh start. In five or at most ten years what we do will also need to be replaced, for science will by then have doubled our knowledge of the nature of life."

In undertaking its task, the BSCS decided that from the very beginning able high school teachers should be selected to work with the biologists from colleges and universities. Often a teacher was paired with a college biologist to work on the same assignment. The teachers' practical knowledge of the difficulties in organizing and teaching high school biology and their firsthand recognition of the verbal and cognitive levels at which materials must be prepared for students of a particular age and background proved invaluable. Just as valuable, we found, was the generation of a feeling on the part of the high school teachers and supervisors that the BSCS was really their own program. The widespread tryouts in selected school systems, the steady flow of critical feedback and suggestion to our writing teams, and the ultimate acceptance of BSCS programs by the secondary school community were attributable in great measure to this policy.

The second major tenet of BSCS thinking developed early in the course of the discussions held by the first steering committee, in 1959. We became convinced that much violence is done in the schoolroom to the true spirit and nature of science because of the textbook writer's apparent assumption of omniscience and infallibility and the teacher's inclination to do likewise. We were of course not the first to arrive at such a conclusion. It has often been observed that teachers who are ill prepared in their subject matter tend to teach by the book. As I have said elsewhere:

> This is one thing that should be anathema in the sciences, which if anything at all have endeavored to dispossess authoritarianism and to substitute for it direct, confirmable observation. Yet legions of our science textbooks serve up to hapless students a crystallized, anonymous science that seems to have descended perfect, like the divine city out of heaven, straight from unquestionable authority. How can we make of any science such a travesty as to teach it upon the word of authority? [23]
>
> To understand science one must see a problem unfold from its beginnings, see progress impeded by traditional ways of thought, learn that scientists make mistakes as well as achieve successes, and observe what experiments brought illumination, and why. One must ask continually, What is the evidence? One must observe how frequently the truth of today is a synthesis of opposing counterviews and countertheories held in their time

[23] Glass, *Science and Liberal Education,* pp. 67–68.

to be irreconcilable. And one must learn from the study of cases how varied and refractory to definition are the methods of science. As to its spirit, there is little of that in either the conventional textbook or lecture. One meets it better in *Arrowsmith* or the *Life of Pasteur*. It is born by contagion; its home is the laboratory, the observatory, or the field, wherever the inexperienced person can observe experience, and the novitiate partake of the zest of discovery.[24]

From opposition to authoritarian teaching in science, the BSCS thus came to a decision not to supply a single authoritative new textbook which might increase the tendency of inexperienced teachers to follow the book blindly. Instead, let there be a choice of three different versions of biology. All of these, we hoped, would be equally sound and equally stimulating. They would, however, adopt different approaches to the subject and would organize it differently. One of them would use an approach commonly used in some of the better college courses, treating first the unity of life in its chemical and cellular aspects, then discussing the diversity of the earth's living beings, from microorganisms to plants and animals, and finally emphasizing the continuity of life through heredity and evolution. This became known as the Yellow Version.[25] The Green Version[26] was ecological in approach and quite different from existing high school biology books. From a consideration of the web of life, populations, and communities, it went on to survey the patterns of life in various types of environment, and in the past as well as present; it looked more briefly at the form and structure, repro-

[24] *Ibid.*, pp. 62–63.

[25] Biological Sciences Curriculum Study, *Biological Science: An Inquiry into Life* (New York: Harcourt, Brace & World, 1963).

[26] Biological Sciences Curriculum Study, *High School Biology, BSCS Green Version* (Chicago: Rand McNally & Company, 1963).

duction, and heredity of the individual; it emphasized adaptation in evolution and in behavior in order to place man in his proper relationship to the biosphere. The Blue Version[27] planned a daring attempt to introduce the high school student to molecular biology by way of the successive stages in the evolutionary development of life on earth. The evolution of the cell, the evolution of the multicellular organism, and the evolution of higher levels of organization—populations, societies, and communities—supplied a frankly evolutionary theme throughout. We have been asked so often why the BSCS decided to produce three distinct textbooks, instead of concentrating upon one, that it is well to state what the real reason was. No one would be left with the delusion that we thought there is only one right way to organize and present the wide-ranging diversified subject matter of biology if we presented the public with three choices and said, in effect: any one of these is as good as any other! Accuracy and faithfulness to the most valid present concepts of biology we would strive to maintain, but an authoritarian approach we would eschew.

The third cardinal principle guiding us has already been stated implicitly. We sought to encourage the teaching of science as a process of investigation and inquiry, rather than simply and wholly as a body of knowledge, organized into concepts and conceptual schemes. John Dewey's opinion on this matter has already been quoted. It is worth recalling that even earlier, in 1899, he had said:

> It is our present education which is highly specialized, one-sided and narrow. It is an education dominated almost entirely

[27] Biological Sciences Curriculum Study, *Biological Science: Molecules to Man* (Boston: Houghton Mifflin Company, 1963).

by the medieval conception of learning. It is something which appeals for the most part simply to the intellectual aspect of our natures, our desire to learn, to accumulate information, and to get control of the symbols of learning; not to our impulses and tendencies to make, to do, to create, to produce, whether in the form of utility or art.[28]

Profoundly convinced of the need of the individual child to learn about science by doing, by gradually broadening his understanding of the methods of investigation through a personal use of them, we debated what biology teaching in the laboratory and field ought to be. We were horrified at the workbooks we examined. Rarely did they depart from the level of questions about the names of structures observed and their presumed functions. The teaching was almost wholly rote—students looked up the names in their textbooks and diligently wrote them down in the blank spaces provided. In place of such busy-work we proposed to substitute, in so far as it might be possible for novices, genuine experimental work involving real unknowns. To accompany each version there would be designed a laboratory program of one hundred or more exercises, each suitable for performance in a single class period. The texts would be tied to the laboratory program by explicit cross reference, and the results of laboratory exercises would be utilized in further development of the text. However, because such a program of short exercises paralleling the comprehensive organization of a textbook might in fact be too disjointed and insufficient in depth to give a student much real acquaintance with scientific investigation, we also designed a dozen "laboratory blocks," each of which could be substituted for six to eight weeks of class

[28] Dewey, *The School and Society*. Reprinted in Dworkin (ed.), *Dewey on Education*, p. 47.

work at an appropriate time during the school year. We would thus emphasize on the one hand that it is not necessary to cover any text or syllabus in its entirety, and on the other hand that a laboratory exploration of some particular area in depth might offer many more opportunities for genuine scientific investigation. At many points in these laboratory blocks the student could find "open-ended" investigation to pursue on his own.

The study of science as a process of investigation and inquiry was further strengthened by the inclusion in the *Biology Teachers' Handbook*[29] of fifty "Invitations to Enquiry" prepared by Joseph J. Schwab and his team of co-workers. Each of these "invitations" demonstrated how a particular biological topic could be used in class discussion to clarify a certain aspect of scientific method, such as hypothesis or diverse causation, or to study some quantitative treatment of data. Another team, originally under the direction of Paul Brandwein, prepared four books containing two hundred unsolved biological problems culled from a great many submitted by active biologists and judged by the committee to be suitable for a gifted high school student to undertake independently.[30]

The fourth and last of the working principles developed by the BSCS was the recognition that certain unifying themes should be used, not as the subjects of particular sections of a book or course, but rather as the warp binding the fabric of biology together. The unifying themes were to appear in

[29] Biological Sciences Curriculum Study, Joseph J. Schwab, *Biology Teachers' Handbook* (New York: John Wiley & Sons, 1963).
[30] Biological Sciences Curriculum Study, *Research Problems in Biology: Investigations for Students,* I–IV (New York: Anchor Books, Doubleday & Company, 1963, 1965).

every chapter, underlie every discussion and auxiliary activity, and permeate the thought of writers, teachers, and students, whatever the organisms with which they might be dealing, whatever the level of organization, from molecule to biome, to which they might direct attention. This intention proved most difficult to carry out, but in the end most rewarding. Seven great biological themes were chosen: evolution; diversity of type combined with unity of pattern; genetic continuity; the complementarity of organism and environment; the biological roots of behavior; the complementarity of structure and function; and regulation and homeostasis. To these were added two unifying themes applicable to all science, but taken here in their biological context: biological science as a process of investigation and inquiry; and the history of the great biological concepts and conceptual schemes. Many a teacher and student has told me that for the first time the organization of facts and ideas around these great themes produced a sense of unity and coherence in the teaching and study of biology that had hitherto been lacking.

There are doubtless many defects in the BSCS books and laboratory programs, defects growing out of the magnitude of the undertaking and the impossibility of keeping every objective in mind at all times. Each of the three versions has already gone into a second edition, in which far more than updating has been attempted. Laboratory programs are more closely knit with the discourse, so that it is not so readily possible to teach only from the book and to ignore the processes of scientific investigation and inquiry. Films have been produced to put in visual form a variety of biological problems and observations that may be used to generate hypotheses, which can then be tested by observing additional data in the latter part of the film. An elective twelfth-grade biology course has been produced which is entirely devoted to a study of the interaction of experiments and ideas.[31] Yet it is still possible for some critics to say that the BSCS emphasizes content and subject matter too strongly, to the detriment of the proper study of science as investigation and inquiry, and for other critics to say just the opposite.

## *Reconciliation and Challenge*

The processes of scientific investigation and inquiry and the study of meaningful content are of course neither separate nor incompatible. As Schwab has said:

> The treatment of science as enquiry is not achieved by talk about science or scientific method apart from the content of science. On the contrary, treatment of science as enquiry consists of a treatment of scientific knowledge in terms of its origins in the united activities of the human mind and hand which produce it; it is a means for clarifying and illuminating scientific knowledge.[32]

We ought, then, to select content in order to exemplify the different modes of scientific investigation and to clarify the different types of reasoning involved in scientific inquiry. Content must illustrate the kinds and sources of error as well as the conditions for success. It must stress the limitations of science as well as its scope. Thus the treatment of content and the treatment of process are knit into one.

[31] Biological Sciences Curriculum Study, *Biological Science: Interaction of Experiments and Ideas* (Englewood Cliffs, N.J.: Prentice-Hall, 1965).

[32] Schwab, *The Teaching of Science,* p. 102.

Although I do not think the charge valid that the science curriculum studies of the past decade, when each is examined as a whole rather than as a textbook alone, have given insufficient attention to the process of science in contradistinction to the content of science, there remains always the danger that a teacher may thwart the real thrust of a curriculum change, may accept the textbook without the laboratory and field work expected to accompany it, may continue to teach in an authoritarian style. I believe this danger is closely linked with the rapid obsolescence of the science teacher's knowledge and command of his field.

There remains a sense in which even the investigation-oriented science curriculum and the renewed teacher might fail to deal adequately with science as process. This is the second of the senses in which George Basalla has used the phrase: the sense of science as social process, not simply as investigative process.

To understand the great conceptual scheme of organic evolution, including the evolution of man, may satisfy only one-half of the vision of man's place in the universe if it does not deal explicitly with the powers and the risks, the moral obligations and the ethical consequences incurred when man becomes the creator of new species and undertakes to modify his own nature. A study of the dynamics of population increase and of the theory of the limiting factors that curb exponential increases remains abstract if it is not specifically applied to man's own dilemma on a planet of finite dimensions. Science through its discoveries generates longer life and better health for mankind, and thereby exacerbates the density of population, the plight of the aged, the pollution of air and fouling of water, the spoliation of the land. Nothing that man grasps for as betterment is unalloyed. Our old folk and fairy tales recognized this truth full well. In our latter-day faith in progress it has been forgotten that the reward and the disillusion are intermixed. Should not our science courses avoid both the siren song of progress toward a perfect technological culture and the horrendous fear of unavoidable technological doom? Perhaps a realization of the benefits and disasters produced by the use of DDT to control insect pests and its consequent accumulation to lethal levels in the world-wide environment would convey the lesson. Perhaps a study of the use of drugs to relieve pain and control behavior and the consequent dangers of addiction or personal or fetal damage would engrave themselves even more deeply on the mind. Whatever lesson be chosen—and dozens suggest themselves—it is undeniable that our science curricula have in part failed, even the newest and best of them, to deal sufficiently with the role of science in the making of human culture, with the problems of the present world and the fair or dread vision of the future of man.

The environmental sciences and the study of behavior are now coming to the fore. Each new technological invention or discovery requires a multiple systems analysis in advance of its introduction. We can now see, for example, that a physical analysis of the effects of nuclear weapons is inadequate without a knowledge of the paths of fallout through biological food-chains and through the intricate metabolic routes within each organism. Strontium-90 and cesium-137 are much alike in the amounts produced from an atomic explosion, and have rather similar physical half-lives; but they are utterly different in their metabolic behavior, which depends not only on their chemical properties but also on the age-long evolution of living organisms that leads them to accept strontium like calcium, but makes them indifferent to cesium. In problem after problem we find that only a total analysis of effects—physical, chemical, biological, psychological—can define the risk or militate against the danger. It may be that the most needed new type of agency in all countries is one that would apply the fullest possible systems analysis to each and every new technological discovery prior to its introduction. Why do our science curricula fail to deal with such questions? Why is the future public not acquainted with the grave issues it must face?

The benefits and risks of our scientific and technological developments cannot be appreciated except in the context of sufficient understanding of both the nature of science as process and the nature of science as content. Here investigation, inquiry, and organized knowledge and conceptual schemes go hand in hand. After all, there are two primary functions of science education: the one, the technical or empirical, being to transmit and extend the knowledge requisite to human power; the other, the philosophical, being to develop an understanding of man's place in the universe. The former, which no one has delineated better than Carl Becker in *Progress and Power,* is the basis of all cultural development, the fundamental thread of human history. The latter has enabled man to conquer superstition and fear, to open his eyes to an illimitable vision. If man's hope for progress is not to prove a delusion, if his fear of malign external forces is not to be replaced by an even greater fear of man himself, the study of the sciences must be bent to the task.

We live, it may be said, in a crisis of conflicting values. That is not new for man. Always his own desires have in some measure found themselves opposed to the family, the tribe, or the state. The submergence of the individual in the enormous populations and national cataclysms of our times is a mere climax or aggravation of the unending struggle, at least in Western lands, to prize and safeguard the rights and worth of the individual. The natural sciences have played a great part in this historic drama. As John Dewey said, already in 1903, "Science has become incarnate in our immediate attitude toward the world about us, and is embodied in that world itself;" [33] and even more pertinently, "Any scene of action which is social is *also* cosmic or physical. It is also biological. Hence the absolute impossibility of ruling out the physical and biological sciences from bearing upon ethical science." [34] It is for that reason, and in the hope that the sciences may contribute to the resolution of our crisis of values, that I have attempted in my book *Science and Ethical Values* [35] to sketch the ideas of an evolutionist about the origins and inevitable conflicts of some human values. There may yet be hope for mankind, since from the perspective of a geneticist the present conflicts of human racial groups appear evanescent, and from the perspective of a psychologist the motivation of nations to war seems conquerable. What is necessary is insight, and insight may come through learning, if we study the right things in the right way.

In her penetrating book *Education for a New Morality,* Agnes Meyer asked the pregnant question: "How, then, do

[33] John Dewey, *Logical Conditions of a Scientific Treatment of Morality* (Chicago: University of Chicago Press, 1903). Reprinted in *John Dewey on Education,* p. 57.

[34] *Ibid.,* p. 55.

[35] Bentley Glass, *Science and Ethical Values* (Chapel Hill, N.C.: University of North Carolina Press, 1965).

we persuade those who are now hostile to science that it is not only essential to the very survival of our nation but also to the development among our people of a truly democratic philosophy?" She answers the question in terms of process: "The very fact that science is a process whereby our ideas become accessible to others points to its fundamental value in a world rent by violent dissensions,—namely, its value as the greatest single binding force among men and therefore a source of international amity."[36] She answers it also in terms of content: *"This is the central challenge to our educators—that subject matter is important only as it opens up new vistas of thought."*[37] Like myself, she acknowledges her great debt to John Dewey in reaching these conclusions. He was a true seer, often misinterpreted, like most seers and prophets in every age. In the reconciliation of content and process in the science curriculum that remains to be fully born he will find his justification. Our world indeed may hang upon a true appreciation of his words.

[36] Agnes E. Meyer, *Education for a New Morality* (New York: The Macmillan Company, 1957), pp. 23–24.
[37] *Ibid.*, p. 59.

# 2

# THE OBSOLESCENCE OF EDUCATION

## *The Growing Demands upon Education*

Scientific knowledge has been increasing exponentially for the past three centuries and a half, with a doubling time less than half of that of the most rapidly exploding human populations. In the United States the profession composed of scientists and technologists is the most rapidly growing of all professions, and will soon outstrip every other single profession in absolute numbers. It can be affirmed unequivocally that the amount of scientific knowledge available at the end of one's life will be about one hundred times what it was when he was born. In rapidly advancing fields of biology, for example, textbooks are scarcely written and printed before they are sadly out of date. The basic understanding of the nature of life has been enlarged more in the past fifteen years than in all of the years before that time. Some 20,000 biological journals publish about half a million reports of original research annually, and the flood of new findings is beyond the power of the most up-to-date computers and indexing services to organize and store them in a retrievable form.[1]

This exponential increase of scientific knowledge has the most profound effect not only upon science education throughout the world, but indirectly upon all other subjects of study, since the world's industrial, economic, political, and cultural life is altered with similarly increasing speed by the translation of scientific discovery into technology. The process alters rapidly and profoundly both what knowledge is most applicable to human needs and what knowledge, from the standpoint of scientific understanding itself, it is most important to teach. It demands constant attention to the revision of the curriculum. It requires extensive alterations of the programs for training new teachers. It makes necessary the continuous retraining of all teachers. It lengthens the period required for adequate training of research personnel and makes postdoctoral training a necessity. It thus contributes to the "brain drain" from the less-developed nations to those of a higher economic and scientific level of advancement, while at the same time the need for vigorous, up-to-date leadership in each country is multiplied through the obsolescence of the existing cadre of university and school teachers. These current consequences of the explosion of knowledge are by no means limited to the natural sciences. They are merely more acute in those areas.

Directly or indirectly the more obvious and most severe

[1] See Bentley Glass, "Information Crisis in Biology," *Bulletin of the Atomic Scientists,* 18 (October 1962): 6–12.

crises of human society are traceable to the explosion of science and technology. The enormous, seemingly uncontrollable expansions of human populations in Asia, South America, and Africa, at rates of 2 per cent and even 3 per cent per annum, clearly result from the reduction of infant mortality through improved hygiene and medical care, as well as better nutrition, in the absence of any parallel reduction in the birth rate. The discoveries of wonder drugs, antibiotics, and vitamins provide only the last chapter of the medical advances that have brought this social crisis upon us. Pollution of the atmosphere by the noxious products of industrial processes and of the internal combustion engines of the ever-expanding hordes of automotive vehicles is now worldwide. Pollution of the earth's waters is reaching disastrous proportions, both on account of the quantities of sewage and wastes produced by teeming populations and because of the effluents of industry and the run-off from lands sprayed with pesticides and herbicides or supplied with mineral fertilizers. Great increases in the standard of living, made possible through science and technology, increase the envy—and ultimately the enmity—of the have-not nations for the fortunate, and especially exacerbate the rage and despair of those depressed, underprivileged classes and colors of people who live in the midst of an affluent society without possessing its fruits. The vast urban dilemmas resulting from an uncontrolled distribution of industry, business, residence, and traffic; the clashes between competing forms of economic and social organizations in the new world created by a scientific technology; and the overwhelming, unremitting threat of nuclear war and the destruction of civilization—these too are product and byproduct of our boasted gains in scientific understanding. Even the unrest of countless students upon campuses throughout the world is, whether they know it or not, a rebellion against an education that seems to have forgotten both the timely and the timeless in a quest for the trivial, a rebellion against a social and political organization that seems impotent to protect mankind against the consequences of man's folly.

In this troubled world, education itself must change as rapidly as science and technology, or the clashing gears of our two greatest human social inventions for adaptation and progress will destroy each other. Certain trends in the reform of modern education must be accelerated. Other transformations must be invented. The conservative nature of the educational system and the slowness of social change, which must overcome inertia and prejudice, may evoke despair. Certainly far too little is proposed, with far too little daring—and even less is done.

Without claiming omniscience, I believe that a few of the major alterations unequivocally demanded by the new age of scientific man are apparent. First, a truly scientific civilization demands a higher order of education than has ever been recognized in our plans for universal education. Mere literacy is not enough. A high order of broad understanding of the essential interdependence of the natural sciences, social sciences, arts and humanities is necessary on the part of the intelligent leadership of each nation. The present trend toward an estimated 7 million young Americans in college in 1970 represents an enrollment of about 50 per cent of persons aged eighteen to twenty-one years. Surely not less than half of any population possess the mental gifts to profit from a college education, and the nation will need that level of general understanding and wisdom if it is to survive. We should then expect that in the year 2000, when our own population is almost certain to exceed a total of 300 million persons, the college population will exceed 10 million. If the trend toward an even higher proportion of college students in the age cohort continues, the number may reach 15 million. It seems quite certain that our college and university facilities must be doubled again before the end of the century! The need for college teachers will rise to exceed 1 million, or over 30,000 new college teachers per year. Since about half of the doctoral production of our graduate schools does not enter teaching, the total annual production of Ph. D.'s should be over 60,000, or twice what is now projected for the year 1975.

The substance of education, like the fund of knowledge upon which it is based and the nature of the society it serves, must change at an exponential rate. In other words, curricula must undergo continuous, radical revision. Significant new knowledge, especially new fundamental concepts and approaches, must be quickly introduced, necessarily at the expense of less important older information that must be pruned away. As for the science component of the curriculum, the emphasis should fall even more strongly upon the development of an understanding of the nature and processes of scientific inquiry and discovery, of methods of verification and experimental testing, rather than upon accumulated scientific knowledge as such. It is important that students come to realize that a science is a growing, changing body of knowledge, not a set of authoritative dicta. For the general citizen, as distinguished from the practicing scientist, a knowledge of the place of science in man's culture and technology, and of its influence upon his philosophy of life, are even more important than a command of scientific facts and principles. Moreover, a good curriculum is adapted to the local scene: the local environment, the local socioeconomic culture, and the local educational system.

## *Educational Obsolescence and the Remedy*

The obsolescence of education in rapidly developing fields of knowledge has become about equal in rate to the obsolescence of an automobile. In five to seven years it is due for a complete replacement. It follows that our times, to a degree generally quite unrecognized, demand a major reconstitution of the educational process, which must become one of lifelong renewal. Perhaps a month out of every year, or three months every third year, might be an acceptable new pattern. Indeed, instead of cramming the educational years of life into adolescence and early maturity, a more efficient plan might be to interrupt education with work periods after elementary school, high school, and college. In any case, programs of continuing education for all professional people must become mandatory, and the educational effort and expenditures must be expanded by at least a third to permit adequate retraining and reeducation.

The obsolescence of education differs only in degree in the several sciences, and even in the humanities. I recently heard a distinguished historian, who had been in university administration for only five years, say that his field had advanced so much in the interim that it would take him a year of concentrated study just to catch up with its development. In spite of a very general recognition of these hard facts, little is done to alter our pattern of education so as to cope with them. Physicians continue to practice medicine although fifty years have elapsed since their youthful preparation. Dentists do the same. Lawyers and engineers live comfortably on their antiquated stock in trade. Teachers slide steadily downhill in their grasp of new developments in their own subjects. What is perhaps worse than any of these is the virtually universal ignorance, on the part of educated men and women, of any advancement of knowledge outside their own professional specialties. Surely we need a complete and thoroughgoing change in attitude toward "adult education," a careful planning of programs and courses appropriately designed for the intelligent adult who has become out of touch with his new world, and a mandatory, cyclic renewal of training for the professional specialist.

The cry of the student throughout the land is for "relevance" in the curriculum. What does he mean? Change has become the major feature of human civilization, driven by an increasingly rapid development of enormous powers to modify and control raw nature; and the advance of science is the principal factor in this technological revolution. Education must prepare each person to cope with changes that are unpredictable. As Heraclitus wisely said: "No man steps into the same river twice." The river flows, and the man ages. All is change. Yet the life of an ancient Greek was not so different from the life of Thomas Jefferson as Jefferson's life from ours today. The content of the curriculum should therefore embrace both the timely and the timeless, for topical and social relevance and timeless natural law alike deepen our perspective and assist us to adapt ourselves to altered circumstances. Unfortunately, the timely and the timeless are both of them often displaced by the trivial, or the significance of the two former is left obscure, so that to the young mind they seem to be trivial. This must not be!

What can be done? I believe that the Japanese have pointed the way. During a visit of several weeks in the summer of 1965, I was privileged to observe the work of six of their Science Education Centers, of which there are now thirty-three, located in almost every prefecture and major city of the country. The first of these centers was founded so recently as 1960. A committee of farsighted scientists and educational leaders had realized that drastic and novel steps must be undertaken to remedy the poor training and reduce the educational obsolescence of the science teachers in the schools.[2]

Assisted by the Ministry of Education, which met about half the cost of construction and a third of the annual cost of operation, the Board of Education in each prefecture was encouraged to establish a Science Education Center. Local control of organization and curriculum was thus assured, a system rather surprising in a country with a strong central Ministry of Education and surely of special interest to educators in the United States.

The organization of a typical Science Education Center is well worth describing. In a building planned by the local Board of Education as a part of the school system, there are usually four laboratories accommodating classes of about twenty-four students for physics, chemistry, biology, and earth sciences respectively. Adjacent to each laboratory there is a large room of similar size for the staff of each depart-

[2] For a fuller evaluation of the role and accomplishment of the Japanese Science Education Centers, see Bentley Glass, "The Japanese Science Education Centers," *Science,* 154 (1966): 221–228.

ment and for storage of equipment, preparation of materials, and some research activity, usually directed toward the improvement of teaching. In addition, each building is likely to contain a modest library, perhaps a lecture room for as many as one hundred persons, offices for the administration, and in some cases a dining hall and dormitory rooms for teachers in residence. Besides equipment very like that of a typical Japanese high school, the center may have collections of minerals and fossils, a greenhouse, a small animal collection, and pieces of expensive equipment such as a donated or second-hand electron microscope, X-ray machine, or 15cm-telescope, which the teachers can learn to use and demonstrate and to which they can later bring their own classes for study and observation.

The most remarkable aspect of the centers lies in their staffing and programming. Each center has a permanent staff, generally composed of one scientist and one or more highly experienced high school teachers for each discipline represented. These staff members plan and prepare, as well as teach, the courses of the center. There are two principal kinds of courses. Short courses, five days long, are given in intensive fashion, utilizing the best of new materials and methods. I was surprised to find that the staff members were very well acquainted with our American science curriculum studies and were making effective, and indeed critical, use of them. The majority of teachers in the local school system, embracing often upwards of sixty schools, are thus provided with effective stimulation and renewal of training. Because teachers in Japan are employed on a twelve-month annual basis, they are subject to duty the year round, and the Science Education Centers are never empty. Summer institutes are numerous. Nevertheless, during the school term, to the maximum extent possible, teachers are released from regular duties to attend the week-long courses while their places are taken by substitutes or other arrangements are made for the students by flexible scheduling. A second type of course is one of much greater scope, occupying a group of twenty-five teachers a full half-year on leave from teaching duties. The depth of training possible in a course running all day, five days a week, for a half year is indeed excellent, and explains the fact that I saw here the best demonstration teaching with high school students I have ever been fortunate enough to observe. The product of such courses is a group of master teachers who are dispersed to different schools and who effectively revitalize there the teaching in their subjects.

The effects of this relatively recent educational experiment have been so dramatic that the Ministry of Education has strongly encouraged the development of new centers. By 1965 the results were so extraordinary that it was decided without further delay to extend the scope of the centers to include mathematics, social studies, languages, and other disciplines.

In the United States, where local and state Boards of Education are so fearful of federal interference and domination, the pattern established by the Japanese Science Education Centers offers particular encouragement. Why cannot our universities, which have hitherto so timorously ventured into the vital task of continuing education, begin now to cooperate with local school systems to establish regional education centers that are far more than hotels or conference centers—that instead will answer the serious challenge posed by educational obsolescence to our entire social fabric? Federal assistance is very well, but it is quite evident that our summer institutes, even though now in existence for more than a decade, skim off only the cream of the teaching profession, and leave the majority of teachers unaided—in fact, more obsolete in knowledge of their subjects and of improved ways of teaching with every passing year. An even more serious indictment of the summer institutes and of present continuing education programs is that, with the rarest exceptions, no serious, continuous thought and effort have been given to design courses really suited to the needs of the teacher who is afflicted by the obsolescence of his early training. It is self-evident that the kind of course well suited for a freshman or sophomore college student is not at all suitable for a teacher who has devoted considerable study to a particular subject but who now stands in need of review and freshening and especially of an adequate explanation of significant new developments and their relationships to the older materials. Nor is the usual advanced course, with its emphasis on narrow specialization, at all useful for a renewal of training and an extension of knowledge.

During the past decade we have seen, in the United States, the development of numerous curriculum studies, first revitalizing the secondary school teaching of physics, mathematics, chemistry, biology, and earth sciences, and later extended to elementary school science and mathematics. To have participated in one of these curriculum studies, to have contributed to such a program, and to have seen it bear fruit in schools across the country has been one of the greatest joys of an educator's life. Yet one must recognize, sadly, that very few of the teachers now in our schools are prepared to take any new curriculum and utilize it well. Often their enthusiasm outruns their knowledge and their current skill. Moreover, the new curricula, which must at best be regarded as no more than a few steps in the right direction, must be continuously revised to keep pace with the advances of science and the development of fresh human problems. Hence the teachers must be trained not once, but many times over, throughout their professional lives.

## *Education in a Brave New World*

Let me then dream a little and try to forecast what the university of tomorrow may become. If I am right, perhaps not more than half the university's total effort and operating budget will be spent on preservice education, whether for teachers or doctors or engineers or other professional people. A gigantic new aspect of the university will be its devotion to the periodic renewal of education, in no haphazard, "weak-sister" way, but indeed as a serious, and for many professions mandatory, cyclical process. The costs of this new phase of education must in large part be borne by the state and the employers, who will learn that profits accrue from continuously reeducated personnel. In respect to the schools, support of such renewal programs must be rendered by the combined efforts of the university, the local school systems, the state, and the federal government.

To make of this enterprise itself a continuously developing, experimental part of the educational system, I can see three new types of organized effort. One of these, as at my own branch of the State University of New York at Stony Brook, we may call an Instructional Resources Center. Here instructional design, use of visual methods and audio tapes, computer-assisted instruction, and other novel experimental educational methods may be combined with facilities for scientific testing of the reliability and validity of such methods. A second organized unit we may call a Center for Curriculum Development. This is an institute in which task forces of scholars and scientists together with experienced teachers give continuing thought to the design of new educational programs, especially those particularly needed for the renewal of education. In this center the advanced methods of the new educational technology are to be brought together with the significant new developments of subject matter, and continuously new syntheses are evolved for actual teaching. No such program would be worth much without the final practical test in the classroom and laboratory. The third organized unit, a Center for Continuing Education, is therefore necessary not only to satisfy the needs for the renewal of education of professional groups and of adults in general, but also to serve as the site where the new methods and ideas are to be put to the test. A steady flow of suggestions and criticisms from the staff and the students in the Center for Continuing Education will provide invaluable feedback for the Center for Curriculum Development and the Instructional Resources Center. This should be a true educational cybernetic system. Here may well be born the fresh, vital influence that will ultimately sweep away the timeworn methods of instruction and the moldy subject matter that so sadly characterize much of our university teaching.

The greatest challenge which education must face in our time is that of coping with the rapidity of change in science, in technology, in human power, in the conditions of man's life. The crux of modern education lies precisely here: that the educated man of yesterday is the maladjusted, uneducated man of today and the culturally illiterate misfit of tomorrow. Education must clearly become a continuing process of renewal.

## 3

# SCIENCE, EDUCATION, AND SOCIETY

THE OVERWHELMING GROWTH of science in our century and the unceasing, dramatic changes in our technology produce inevitable alterations in the content of science as taught in school and university. The timeless truth expands in all directions; the timely reality takes on new shape and structure.

There is, however, a complementary aspect that is fully as important. What is taught in school and university supplies the concepts, the principles, the paradigms and conceptual schemes that in large measure shape the science of the future. An infinitude of data that might be collected are not; millions of experiments that might be done are never attempted. Always the scientist exercises some subjective selection in his activity, even in the basic acts of observation; for what he observes and what he declines to observe, consciously or unconsciously, is the fruit of his whole intellectual development, the conditioning of his training, the bias of his current frame of thought. In a very real sense, then, the future of science must grow and develop, like a living organism, out of each preceding phase. It is this that makes it possible to predict, in some measure, the future course of science.

Yet it is possible, of course, only in modest degree. The future of science remains ultimately unpredictable because of the genius, the maverick who can discard at least once in a lifetime the constraints of the currently accepted conceptual model, break through the bounds of the paradigm, and in consequence perform the totally unexpected experiment, formulate the new theory that displaces the old, unite in harmony such contradictory views as wave and particle concepts of light, or epigenetic and preformist views of the origin and development of the organism. This genius is nourished by the recognition that the old conceptual scheme is subtly but increasingly inadequate. He rebels against the current scientific dogma upon which so many current studies are heuristically based. Though later recognized as the creative genius of science, he is first its heretic and destroyer.

A major problem in the study of science as a social process is to find a secure way to distinguish the creative genius, at first appearance, from the crackpot. Another significant problem is that of science education: how can we nurture and develop, instead of stifling, this embryonic spirit that accepts the current scientific world view only to become dissatisfied, that breaks through the trammels of scientific convention to scale new heights? To emphasize that science is a process of investigation to be pursued actively by the student, rather than simply a more and more crystallized body of facts and established concepts or laws, to be comprehended or memorized—that will certainly help. It is probably not enough. In addition, I believe, the study of science as a social process, beginning in the past but sweeping impetuously and comprehensively into the future, will likewise serve to place science, man's chief tool, his noblest work of art, as Carl Becker would say, in a fuller perspective.

In that belief, the third of these essays will be devoted to a sketch of the characteristics of the growth of science during the past century and a half, the limiting factors that even now are beginning to come into play, the urgent need for technological assessment and foresight, and the consequent intimate relation of science in the twenty-first century to the nature of education at that time.

## *The World in 1820*

1820! Europe was painfully collecting itself after the Napoleonic debacle. In the United States, a new state a year was being carved from the Louisiana Purchase, and President Monroe was considering how best to keep the European powers from intruding further into the New World. India was at length fully in British hands. Australia had a few scattered convict colonies on the coast. Africa, the great unknown continent, was penetrated mainly by Arab slave traders. Even its coasts, except at the Cape of Good Hope, were not yet carved into the colonies that would shape the political map of the world during the nineteenth century. China and Japan still slumbered behind doors closed to the Western world.

Travel was scarcely more rapid than in the days of the Romans and still depended mainly upon horse, coach, and sail.

It required six days to make the trip from New York to Washington. Even a letter carried by postal relays took three days. The first railroads were being built in England, but the steam locomotive was in its infancy, and the stagecoach was as rapid if not as economical. Highway construction had not improved from Roman times. The first steamship to cross the Atlantic Ocean, the *Savannah,* in 1819, used only auxiliary steam power and portable paddles to supplement her sails, and took twenty-nine days eleven hours for the crossing. A fast sailing vessel, all wooden, at the same period required twenty days, although the record was thirteen days seven hours. There were of course no automobiles or airplanes.

Communication was as slow as human travel. There was no telegraph, no telephone, no radio. Except for the early use of steam engines in the British mills and mines for turning machinery and pumping water, the sources of power had not changed from ancient times. The windmill, the water wheel, and human and animal labor supplied the basic needs. There was of course no electric power or lighting. Illumination at night still depended upon natural oils, the olive oil of the ancients or whale oil from the fisheries. Household heating was shifting from open coal or wood fires to use of the more efficient iron stove, such as the Franklin stove. Although man in northern climes thus tempered the rigors of the winter cold, no one as yet dreamed of attempting to produce artificial coolness during the heat and humidity of the summer.

Sanitation was no better than in Roman times, and probably nearly everywhere considerably worse. No one, of course, connected the appallingly high infant mortality or the frequent plagues and epidemics which occurred with the spread of germs or parasites. Malaria, the greatest of human killers, was not associated in men's minds with mosquitoes, nor plague with rats and fleas. Only direct contact with diseased persons was recognized as being important to avoid. Nutritional deficiencies commonly went unrecognized as such, although Sir James Cook and others had discovered the importance of fresh fruit in preventing scurvy among sailors. In Spain the average life expectancy was under thirty years; in England and the United States it was under forty years, and was not to begin its phenomenal climb until the second half of the nineteenth century. No immunization was practiced except through vaccination against smallpox, and vaccination was voluntary and sporadic, rather than required. Surgery was necessarily without anesthesia. Since there was also neither antiseptic nor aseptic surgery, most operations ended in terrible infections followed by death, except for amputations which could be followed by cautery.

The population of the world had reached about one billion persons. It was growing with particular rapidity in Europe. In twenty years, since the first English census, the population of that country had increased by more than 50 per cent. The gloomy doctrine of Malthus was widely accepted.

Science in 1820 was virtually limited to Western Europe, with minor participation in Russia, the United States, and elsewhere. In all the world there were probably fewer than a thousand persons who thought of themselves as natural scientists. There were no regular scientific journals, such as *Nature* or *Science*. The proceedings of the scientific academies sufficed for publication of new discoveries and reports, or these were incorporated in books. Physics lacked an electromagnetic theory and a conservation of energy principle. Radioactivity was unknown. The spectroscope had not been invented, and spectrum analysis was primitive. In chemistry the study of the carbon compounds was getting under way. The atomic theory was new. Some forty elements were known. There was no family grouping of these—no periodic table. Geology was still explaining the succession of strata and the preservation of fossils on the basis of catastrophes of flood or fire. No one recognized the immense age of the earth or the formation of the strata through the ordinary processes of erosion and sedimentation, for Lyell's great work was still a decade in the future. In astronomy Uranus had been discovered, but Neptune not yet. No star distances were known. The universe still seemed comfortably small, for definite proof was lacking that spiral nebulae are extragalactic objects enormously numerous and equaling or exceeding our own galaxy in size. That indispensable adjunct of astronomical observation, photography, was in its earliest phase of invention.

The biological sciences were even less mature than the physical sciences, being still largely observational rather than experimental. The very word "biology" was only a decade old. There was no cell theory, no germ theory of disease, no Mendelian genetics. That sexual reproduction leads to the origin of new individuals through the fusion of an egg cell with a sperm cell had been guessed at, but not established. Biochemistry was in its first infancy. No enzyme had as yet been found. Proteins, identifiable by their nitrogen, were recognized to be of general importance; but nucleic acids were not to be discovered for another half century. There were scientists who believed in the gradual evolution of species to their present forms, but no convincing summation of evidence had been made, and above all no one had proposed a satisfactory theory to account for the process of change and its product, the ingeniously adapted structures and activities of living things that suit them so admirably to their environmental situations and their ways of life. These were still universally attributed to the Divine Wisdom of Providence. A science of behavior, including human behavior (psychology), could not be said to exist at all.

Perhaps mathematics alone was, relative to its present status, more mature and more advanced in 1820 than the natural sciences, for the great developments of the seventeenth, eighteenth, and early nineteenth centuries seem to have left but minor areas for exploration. Gauss had already made significant advances in the study of probability and error, number theory, and the analysis of non-Euclidean space. True, others would greatly develop and apply these methods and concepts. To a nonmathematician it seems that perhaps the greatest era was past, and that nothing like the renewal of physics through relativity and quantum theory, or the transformation of biology from an observational to an experimental science using the most sophisticated tools of physics and chemistry to penetrate the ultimate secret of life, has taken place in mathematics in the past 150 years.

### *The Exponential Increase of Science and Technology*

I have sketched briefly the state of the civilized world and of the natural sciences in 1820 because we must continually picture it in our minds if we are to comprehend the enormous difference that has transpired and see graphically what modern education must undertake. Perhaps 150 years is in fact too long a span of time. One might equally well have chosen the year 1900, or even the year 1920, to make the point, for the essential characteristic of the scientific and technological change that is occurring is its exponential rate.

In many ways the year 1920 is more similar to the year 1820 than to the year 1970. Take, for example, the world population as one measure. In 1820, about 1 billion; in 1920, an estimated 1.8 billion; in 1970, an estimated 3.6 billion. In the first century, almost a doubling (80 per cent gain); in the following *half* century, a full doubling. Take the average life expectancy in the United States as another measure. In 1820 it was about thirty-nine years, in 1920 it was about fifty-four years, in 1970 it will be approximately seventy-one years. The first century's gain was fifteen years in life expectancy; the recent *half* century's gain was seventeen years. Those figures represent the social applications of sulfa drugs and antibiotics, pure milk and water, and universal vitamins to the reduction of infant and childhood mortality. The mortality from tuberculosis was cut by more than 90 per cent during the 1950's, following the introduction of isioniazid supplemented by streptomycin.

As for transportation, communication, and entertainment, why labor the obvious? In 1920 we had no commercial air travel, no speedy and reliable long-distance telephone service, no radio broadcasting, no television, no movies with sound or color.

The increase in scientific and technological work has likewise been exponential, and at an even faster rate. In a penetrating discussion of these matters, Derek J. de Solla Price has documented the increase in the number of scientific journals since 1665.[1] From about 1750 to the most recent time

[1] Derek J. de Solla Price, "Diseases of Science," in *Science since Babylon* (New Haven: Yale University Press, 1961), pp. 92–124. Also "Prologue to a Science of Science," in *Little Science, Big Science* (New York: Columbia University Press, 1963), pp. 1–32.

the number of journals has increased by a factor of ten in every half century, that is, with a doubling time of about fifteen years, with scarcely a perceptible deviation from expectation. This increase is by a factor of a thousand for a century and a half, and for the most recent fifty years an absolute increase from approximately 30,000 journals to about 300,000. Price further points out that abstract journals came into existence about 1830, when the number of journals reached a total of 300, and it was consequently no longer feasible for any polymath among scientists to read and digest them all; hence a summary of journal contents became necessary. Since that date the increase in the number of abstracting journals has followed exactly the same rule of increase as the primary journals—a tenfold increase in every half century! Their number has now also reached the critical number of 300, and there now arises a clamor for abstracts of abstracts. Electronic sorting of abstracts by computer, Price thinks, will not fully serve the purpose, since it may at best produce a tenfold compression, whereas one of 300-fold is needed.

The increases in the number of scientific journals and of abstracting journals of course reflect the increase in the number of scientific papers presenting results of original research. *Biological Abstracts,* for example, has increased its annual coverage from 40,000 abstracts per year in 1957 to about 110,000 per year in 1967—and that is admittedly a quite incomplete coverage of the present output in the biological sciences. *Chemical Abstracts* has twice as large an annual coverage (244,000 in 1967). It is probable that at the present time at least 600,000 papers reporting original research are published each year. Moreover, it is quite clear that the exponential relation has been holding for research papers just as for journals. Price has documented this for the field of physics, in which, except for an offset during World War II, the exponential curve is that for a doubling in every twelve years.

The number of original research papers published, of course, reflects the number of working scientists at any given time. Lotka's Law of Productivity states that the number of authors of *n* papers is proportional to $1/n$.[2] The law has been demonstrated to hold with surprising accuracy in a number of studies of productivity. As a consequence, one can calculate that for every three and a half papers there is one existing author.[2] The number of scientists is approximately one-third the number of papers, whether this be figured on a cumulative basis or for a given cross section of time. A million scientific papers published annually thus corresponds to a third of a million working scientists.

Scientists, too, have been increasing exponentially in numbers over the past 150 years, with a doubling time somewhere between ten and fifteen years. Thus, the U.S. Census shows "scientists and technologists" as a major professional group including scientists, engineers, and technicians. In 1870 there were in round numbers 12,000; in 1900, 90,000; in 1930, 450,000; in 1950, 1 million; and in 1960, 1.35 million.[3] The curve fits with remarkably little irregularity the

[2] Price, *Science since Babylon,* p. 106. Also Price, *Little Science, Big Science,* pp. 42–50.

[3] National Manpower Council, *A Policy for Scientific and Professional Manpower* (New York: Columbia University Press, 1953), Table 1, p. 44, supplemented by United States Department of Labor, *Manpower Report of the President, and A Report on Manpower Requirements, Resources, Utilization and Training,* (Washington, D.C.: U.S. Government Printing Office, 1963), Table G6, pp. 202–204.

exponential expectation for a doubling time of thirteen to fourteen years. The scientists proper—those responsible for the production of original research in science—constitute about 11 per cent of the total in this "scientists and technologists" category. If the scientists who teach in colleges and universities are added in, instead of being counted as teachers (as does the census), there are about one-fifth more, or 13 per cent of the augmented total.

The professional category of scientists and technologists now constitutes about 20 per cent of the entire professional group in the labor force. The other categories are teaching; health; religion and social welfare; law; arts, letters, and entertainment; and "others." Teachers are the most numerous profession, but the scientists and technologists are rapidly overtaking them. The scientists and technologists somewhat outnumber those in health and are about equal to the three smaller categories combined. "Other" professions also about equal the scientists and technologists. It seems rather unlikely to me that there will be much further relative gain on the part of the scientists and technologists among these professional groups, because of the high social needs for the others. Perhaps the upper limit for scientists and technologists might be about 25 per cent of all professional workers. Further increases in absolute numbers would therefore have to arise from (1) an increase in the size of the total population, and (2) a possible increase in the proportion of professional persons in the total labor force. The former would not long permit continuation of a faster doubling rate than that of the population itself. At a 1 per cent annual rate of increase, the general population will double in seventy years. Since the total population must in a few decades cease to increase and must become stabilized, no great increase in number of scientists and technologists can be expected from this source beyond one or two additional doublings. Such an increase cannot be continued for more than another century without the gravest consequences resulting from increasing density of population.

Some increase in the proportion of all professional persons in the labor force may seem to offer more hope for a less drastic curtailment in the growth of science. Yet here too there seem to be ineluctable limits. To serve adequately in a professional capacity, the equivalent of a college education is widely regarded as a minimum preparation; but very few persons falling into the lower half of the normal distribution of intelligence have sufficient capacity to complete a college program successfully. In other words, we may set a limit for professional competence to include about 50 per cent of the total labor force. Even to train such a number professionally would require a threefold or fourfold increase in our college and university enrollments, for in the United States we are now graduating only 750,000 a year and not all of these graduates enter the labor force. Even so, the upper limit for the proportion of scientists and technologists in our present forms of society may be about the product of the maximum proportion of scientists among professional persons and the maximum proportion of professional persons in the labor force. This product is ¼ times ½, or ⅛ of the total labor force. To attain even that large a proportion we shall need to reduce greatly the need for ordinary nonprofessional labor by extensive automation and increased efficiency.

At their present rate of expansion, the professional and technical groups have slowly increased from 2.8 per cent of the total labor force in 1870 to over 11 per cent at the present time. Until there can be a speedier increase in this professional component of the labor force, the scientists and technologists are limited to whatever increase in numbers can come from some slight gain in their relative proportion within the professional groups. A doubling during the decade just ending would be barely within the limits imposed by an increase of all professionals to 15 per cent of the labor force and of scientists and technologists to 25 per cent of the professionals. Even an increase of the professionals to 20 per cent of the labor force during the decade 1971–1980 would permit only a 75 per cent increase in the numbers of scientists and technologists. In other words, we are clearly at the flexure of the logistic curve that warns us that exponential expansion must halt.

The clear implication of this line of reasoning is that the *relatively* more rapid increase of scientists and technologists in the population of the United States must soon end. Any increase thereafter can still be great, but at a maximum must keep pace with the total growth of population, which itself is likely in a few decades to reach a saturation point, as was said earlier. Presumably the same conditions will apply to every other country as it reaches the same levels of technological development and total density of population.

Derek Price [4] points out that the "stature" of science—I think we might in fact call it the frontier or attainment of science—doubles more slowly than the actual productivity measured in units of journals, papers, or scientists. It doubles in about three decades. Price attributes this to the cumulative nature of science, which like a pyramid or pile of stones must

[4] Price, *Science since Babylon,* pp. 119–121.

increase in volume as the cube of any gain in height. A doubling in height requires eight times the support. This very suggestive analogy may be approximately correct. If so, it points out a serious difficulty that will face us in the near future. In those nations most highly developed technologically, we have become quite dependent in our economy upon an improvement in technology at a rate of 6 to 7 per cent per annum. I know of no studies that relate the rate of technological improvement quantitatively to the increase of science, but it seems evident that the ratio is greater than unity. Perhaps a 2 per cent per annum growth in science can support an 8 per cent technological advance. In that case the doubling time would be about right—thirty-five years. But if the 2 per cent per annum advance required is not in the total quantity of scientific work but in the stature of science, of the thrust of the growing points at the frontier, then actually only an 8 per cent increase in the annual scientific productivity might suffice to guarantee an 8 per cent technological advance. Within the next two decades an 8 per cent annual increase in science is clearly going to move beyond our reach.

I have not yet considered the cost of doing scientific research. It is of course evident that costs are increasing far more rapidly than either scientific manpower or productivity. Not merely must we cope with inflation and the problems of an expanding economy, amounting to some 3 to 6 per cent per year. Far more serious is the increasing cost that comes from the ever more sophisticated and expensive equipment which scientists in every field must now use if they are to work at the forefront of knowledge. Spokesmen of science have optimistically advocated a 15 per cent per annum increase in the support of basic science as being needed to match inflationary costs and to equip and support the growing scientific manpower. Yet such an increase is at an exponential rate greater than any cited in this analysis so far. It plainly could not continue for more than a few doubling times of 4.7 years without consuming the entire Gross National Product. The two billion dollars of current annual support of basic science in the United States would in twenty-four years equal the present defense budget (Vietnam War included) and in forty years would exceed the Gross National Product. Evidently we are limited to a very brief growth at any such rate. What seems so clearly to be needed is a full-scale study of the relation of growth in science to technological improvement, to population size and rate of population increase, and to the balanced welfare of other occupations.

## *The Limiting Factors*

Studies of population growth show that the logistic curve followed by populations is one of exponential growth with saturation. Inevitably, in a world of finite resources and finite space, limiting factors sooner or later must curb the logarithmic phase of growth, and it gradually diminishes in rate until a maximum, or asymptote, is reached. (Figure 3–1). Several types of situations may follow. The asymptotic level may be maintained indefinitely, provided the limiting conditions merely restrict further increase.

If a particular limiting factor is not absolute, like the finite extent of space, then upon relaxation of the limitation the population may resume exponential growth until again some limiting factor curbs it. This relaxation of limitation may even occur repeatedly, so that a stepped pattern of alternat-

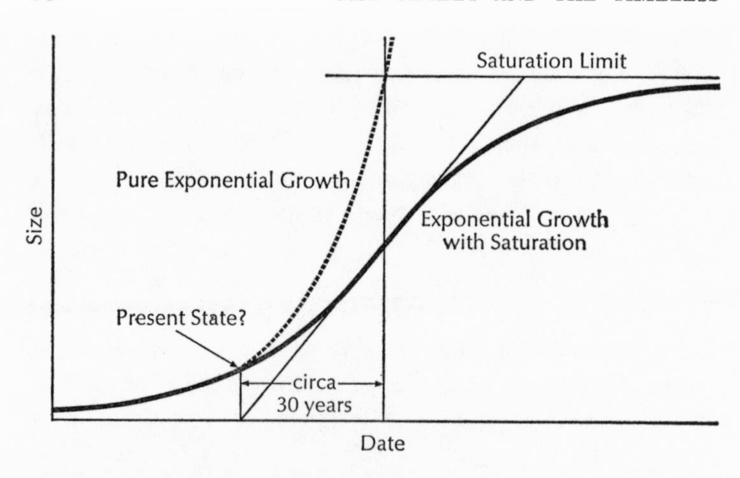


**Figure 3–1. General Form of the Logistic Curve**

After Derek J. de Solla Price, *Science since Babylon* (New Haven: Yale University Press, 1961).

ing periods of exponential growth and equilibrium arises (Figure 3–2a). More commonly, there is a negative feedback from the presence of maximum numbers. In a limited space, the waste products of the organisms will tend to accumulate and may exert a toxic effect upon the members of the population. In this case, the population will decline in numbers, and if the environment is not purified, may eventually become extinct. In more complex situations, where the environment may become purified through a recycling process and where growth of the food supply is enhanced as the number of feeding organisms diminishes, an oscillating equilibrium may become established (Figure 3–2c, d). Complex, alternating, oscillating cycles are quite frequent when the numbers of a predator species and of its prey are examined together. The maxima for the predator lag behind the maxima of the prey species, but as the numbers of the prey diminish, those of predator necessarily decline also (Figure 3–3).

If the growth of science behaves like a growing population of animals or microorganisms, as the evidence to date so clearly indicates it does, we should be interested in the effort to determine the limiting factors and to forecast the extinction or type of equilibrium most likely to occur. I believe there are at least five major limiting factors well worthy of study, but it is quite impossible, in the absence of any critical investigation of the action and interplay of these factors, to say now what may ensue.

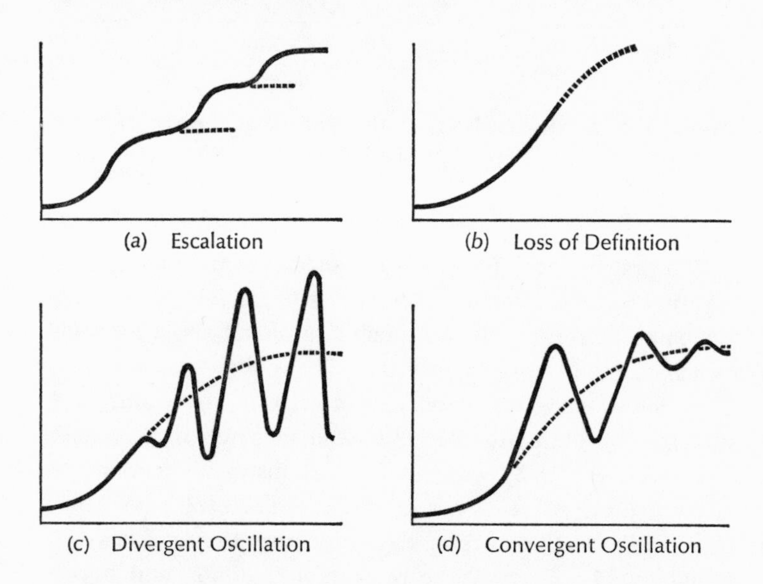


**Figure 3–2. Ways in Which Logistic Growth May React to Limiting Conditions**

After Derek J. de Solla Price, *Little Science, Big Science* (New York: Columbia University Press, 1963).

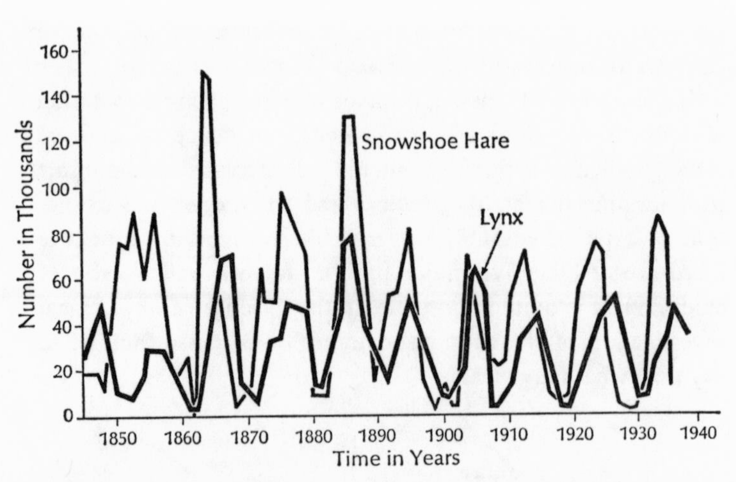


**Figure 3–3. A Comparison of the Estimated Yearly Number of Snowshoe Hares with That of Lynxes.**

After BSCS, *Biological Science: Molecules to Man* (Boston: Houghton Mifflin Company, 1963).

The first of these factors is the sheer volume of scientific information that already exists and is so rapidly accumulating. For a scientist to be adequately informed about even his own specialty is almost impossible. I have been informed that many Soviet scientists spend half of each day in reading and studying the literature. Perhaps a phenomenal speed-reader can cover, especially with the help of abstracts, as many as fifty articles in a half day, at a rate of ten or twelve per hour. If we estimate 250 working days in a year, the total covered would be 12,500 articles, a respectable amount, and nevertheless barely a tenth of the annual output in either biology or chemistry, and less than the output in most active specialties in these fields. What most American scientists do is, I fear, at the opposite extreme. They have frankly surrendered to the impossible and no longer read very much, even in their own fields of work. They do some scanning, on occasion read an important article in full, but for the most part depend upon word of mouth for guidance to the important new work in their specialties. Hence arise the "invisible colleges," which meet annually or semi-annually, and membership in which is informal but open only to the elect, since the special purpose of such colloquia is to enable free and roving discussion to occur within a group of modest size.[5] One of my own studies, in 1958, revealed that most of my scientific colleagues in all fields learn of the work particularly significant to the development of their own investigations through hearsay or chance.[6] The "invisible college," however, represents an unspoken effort to return to the free communication that prevailed among scientists in the early days of the academies, when every scientist in a country knew every other scientist there, and true colloquia were possible. Today, it is implicitly a policy of exclusion and a tacit admission of defeat. The accustomed channels of publication are too slow and too gorged with old and new information to be used effectively. We are drowning in seas of paper! [7]

The second limiting factor in the growth of science is closely allied to the first. It is the ever increasing specialization of the modern scientist, who must restrict his energies to

[5] Price, "Invisible Colleges and the Affluent Scientific Commuter," in *Little Science, Big Science,* pp. 62–91.

[6] Bentley Glass and Sharon H. Norwood, "How Scientists Actually Learn of Work Important to Them," *Proceedings of the International Conference on Scientific Information,* Area I (1958), pp. 185–187.

[7] Bentley Glass, "Information Crisis in Biology," *Bulletin of the Atomic Scientists,* 18 (October 1962): 6–12.

a manageable area if he is to understand it sufficiently to carry on further successful investigations. This tendency of modern science has long been noted. It becomes steadily intensified as the volume of scientific knowledge increases. I well remember, for example, the latest of the McCollum-Pratt Symposia, on the subject of *Light and Life,*[8] in which physical chemists, biochemists, and biologists were brought together for the purpose of developing a mutual understanding of the role of visible light in the attributes and activities of the living world. But the three groups spoke each a different scientific language, unintelligible to those of the two other groups. Some cross communication no doubt occurred, but the impedance was enormous. Francis Bacon divided scientists into the pioneers, who dig, and the smiths, who refine and hammer.[9] It is all too evident that in the twentieth century we have produced enormous numbers of pioneers, but very few smiths. Yet it is the smiths who construct the great unifying theories, such as the periodic law of the elements or the atomic numbers, the electromagnetic theory, the cell theory, the germ theory of disease or the theory of organic evolution, to name a few of the magnificent achievements of the nineteenth century.

It is well recognized that many, if not most, scientific breakthroughs come about when the techniques and concepts of quite different scientific fields are brought together in an original synthesis of insight and imagination. Yet it seems unavoidable that the increasing narrowness of specialization tends to reduce the probability that this will happen. Too

[8] W. D. McElroy and Bentley Glass (eds.), *Light and Life* (Baltimore: Johns Hopkins Press, 1961).

[9] Francis Bacon, *The Advancement of Learning,* Book Two, VII, 1.

many scientists have blinders over their eyes and can see only straight ahead. What excellent diggers! But where are the smiths, who must refine and hammer, create alloys, fashion a meaningful whole for the young blood coming into science in each generation? As editor of a review journal, the *Quarterly Review of Biology,* I feel considerable fervor and even pain by reason of the steadily increasing difficulty of obtaining broad, critical reviews of the sort needed by working scientists and students alike. The literature is too vast and too disorganized. One's reputation is enhanced more by ten so-called original papers, each digging out one more minuscule fact, than by writing a comprehensive and original review article or book. The young men must get ahead; the old men are too tired. Hence the mountains of mined ore accumulate, as Bacon might have put it, around the mouths of the mines, but nothing is done with most of it. It lies there until it is buried by fresh mountains of ore, and perhaps some day will need to be re-exhumed if it is ever to fit into the constructed edifice of science.

The third limiting factor in the growth of science I take to be the rapid rate of educational obsolescence which is imposed upon us by the growth of science, technology, and change in human society. This condition has already been discussed in the second essay of this book. To recapitulate: in about eight years an education in science must be thoroughly renewed, extended, and reorganized; or its possessor is beyond hope as a practitioner. If applied science and change in the character of human society are subject to a doubling time of a generation, say thirty years, then all curricula should be reformed and all patterns of education remade completely within the same period. Our schools and universities nevertheless continue to plan for an education that is crammed into the first twenty, or at the most twenty-five, years of a person's life. Nothing short of the substitution of lifelong formal educative processes combined with professional work can meet the demands of our new technological society.

The effects of educational desuetude are perhaps most dramatic in the examples of the physician and the teacher. What is necessary for both these groups, and others in like straits, is, first, never to lose the habit of study and learning, and second, to be granted the time and financial support, in the form of frequent, perhaps short, leaves of absence to undertake the necessary renewal of education. I am convinced that the typical "sabbatical" of the university comes at too great an interval for our present rate of technological and social change. I am equally convinced that a regular cycle of mandatory leaves, regarded as an essential part of one's professional duty and paid for as such, is essential. Finally, the colleges and universities must begin to revise their programs so as to provide fully adequate and appropriate renewal of education for professional people, or else some new institution must be devised to fill this gap. Our summer institutes for teachers of the natural sciences or mathematics, for reasons already stated, can scarcely serve. Institutes for engineers and physicians are even more sporadic and ineffectual. The Japanese, as was described in the previous chapter, have done better than most nations in recognizing this need and moving to meet it.

A fourth limiting factor in the growth of science I take to be the long-term, unanticipated side effects of the introduction of insufficiently tested technological developments into the social and political economy. The hope of immediate gain too often clouds our vision of consequences. Even laudable motives may conspire to produce disaster. The conquest of infectious disease leads to an explosive increase in the world's population, because there is no immediate compensatory reduction in the birth rate. More convenient motor transportation produces smog. Denser populations aggravate water pollution and generate psychological stress and unrest.

As the peoples of the world become ever more aware of the destruction of their environment and the compaction of teeming populations into miserable urban masses, as with dismay they recognize the loss of the beauty of nature through the encroachments of human industry and technology, it seems natural that they will turn against the forces they will hold responsible. I predict more and more massive resistance to technological change. I predict in equal measure a growing hostility to science, which will be held accountable, and a dwindling measure of financial and social support. Hence the fifth, and perhaps final, limiting factor in the growth of science—the psychological resistance of and the restricted support supplied by a population inadequately educated in the understanding of science and militantly opposed to it because of its identification with the technological annihilation of the human environment.

## *Agencies for Technological Prognosis and Assessment*

These consequences are only to be avoided by sufficient foresight and effort on the part of the scientific community itself. I have thus come to the conclusion that what each country needs, more than anything else, is a new type of agency —whether governmental or private I would not say—an agency to study as effectively as possible the long-term side effects of new technological developments before they are permitted to be introduced into our complex social organization. Many recent environmental, economic, and medical disasters could have been avoided by applying what knowledge we already possess or by making some further, more extensive tests. There was no need for the thalidomide tragedy. There was no need for the Santa Barbara oil damage to the beaches and ocean life. There is no need for the best agricultural land to be used up for housing developments spreading mile on mile without open spaces for trees and recreation. Man is a social animal living in a complex environment. He ignores the balance between his own populations and the nature and resources of his environment at his mortal peril. Nature's cycles are mighty but often delicately subject to overthrow, and civilizations have before now destroyed themselves through waste, ignorance, and folly. All is now repeated on a far grander, worldwide scale; as man seizes the powers of the gods and wields them he strides forward into the Götterdämmerung.

Many voices are now raised in favor of "technology assessment." Most critics of present ways of altering and managing our technological methods seem, however, to regard the need as merely one of finding a simple palliative measure, like an effective precipitator in a factory smokestack, an afterburner on the exhaust of a car, a straight-chain detergent instead of a branched molecule detergent, or a labile pesticide instead of an enduring one such as DDT or dieldrin. As Barry Commoner has recently stated in a government hearing on a proposed resolution to establish a Senate Select Committee on Technology and the Human Environment,[10] that is not at all

[10] Barry Commoner, Testimony before the Subcommittee on Intergovernmental Relations of the Senate Committee on Government Operations concerning a Resolution to Establish a Senate

the real solution. Part of our critical problem is that our technology is already far too efficient and at fault in major ways. The well-known pollution of Lake Erie beyond all hope of immediate recovery is not wholly attributable to the industrial wastes that pour into it from our lakeside mills, but arises even more extensively from the very efficiency of our modern treatment of sewage, which empties sterile but vast quantities of nitrate and other inorganic fertilizer into the waters of the lake. Like nitrate fertilizer on land, the nitrate in the lake stimulates an overgrowth, or "bloom," of algae that turns the waters into a thick pea-green soup, depletes the oxygen, kills fish by myriads, and produces by its upset of the normal ecology of the lake a sort of green desert.

Our extensive use of mineral fertilizers on land has ensured America a bountiful food supply and has become the indispensable stay of the American farmer. Yet these fertilizers tend also to destroy the organic composition of the soil, after which they run off in great quantity into the streams and rivers and find their way into the lakes and seas, where they add to the polluting effect of our treated sewage. The usual remedy sought by the farmer for the increasing losses of the fertilizer he applies is to add still more. In arid lands, the excess fertilizer seeps down into the soil with the irrigation water, and if an insufficient volume of water is used to carry it away, any subsequent rise in the water table brings it back to the topsoil and converts the land to a saline waste.

A principal effect of our increasing production of smog is the release of large quantities of nitrogen oxides into the air. By reducing, with more efficient exhaust burners, the amounts of hydrocarbons released into the atmosphere, we

Select Committee on Technology and the Human Environment, April 24, 1969 (still unpublished).

eliminate less and less of the nitrogen oxides by chemical reaction, and eventually they too come to earth in the form of nitrate added to the rainfall. In pioneer days, rainwater was as pure as distilled water and was often used as such. It is now heavily charged with mineral content. The increase in the fertilizer content of rain and the increasingly heavy applications of mineral fertilizers to the soil yield in many areas such a superabundance of nitrate that we are beginning to experience cases of nitrate poisoning of infants from eating spinach and other leafy vegetables which concentrate nitrate in their tissues.

Enough has been said to indicate that the most efficient and initially advantageous technological developments set off chain reactions that surprise us with their inimical consequences. Although unpredictable in an earlier stage of environmental science, they can now be more and more fully foreseen. It is, for example, predictable that the large-scale introduction of any technological method that increases the release of nitrates into air, soil, or water will aggravate still further an already critical problem of our civilization. What is obviously needed is a careful study of the total ecological system and the impact upon it of our alterations and intrusions. Especially necessary is an evaluation of the stability, or "homeostasis," of the environment, to apply a physiological term that is highly useful in describing the ability of a system to restore itself after displacement from an equilibrium and to maintain a relatively stable yet dynamic state near the optimum for the maintenance of life.

In visiting the scientific institutes of Czechoslovakia in 1966 I was surprised to discover, among the institutes of the Slovakian Academy of Sciences, one with the intriguing name, "The Institute for the Study of the Biological Landscape." Inquiring about the meaning of this title and the subject of study of the institute, which I later visited, I was told a tale of great interest. After the close of World War II, the economic planners of the country decided to develop Slovakia, previously almost exclusively agricultural, into an industrial land. According to plan, a large aluminum processing plant was to be located somewhere in Slovakia. The military authorities stoutly defended the choice as a site of one of the long, narrow valleys in the Middle or Lower Tatra Mountains. Some biologists deplored this choice of site, and pointed with reason to the probability that smelter fumes would cause considerable damage; but they had no hard facts of experience to submit and were overruled. The plant was constructed in the valley. At that date, prior to the completion of the pipeline that now brings oil from the Caspian Sea to Slovakia, the available fuel was the local Czechoslovakian coal, high in sulfur content. Within a few months after operations opened, it became obvious that the valley had a typical atmospheric inversion layer which trapped the fumes in the valley. The sulfur dioxide content of the air rapidly increased until every living plant or animal in the valley was dead, save for the men in the mills. They had to reside at a great distance, a situation causing much loss of working time, and even at times had to wear gas masks to continue working in the smelter. The industrial result was an almost total disaster, which might have been, but was not, foreseen. Hence arose the belated official recognition of the need for an institute to study the "biological landscape," in order to prevent it from being turned elsewhere into a moon waste.

The foregoing account deals only with one example, on a

limited scale, of the destruction man wreaks upon his environment. Examples of similar disaster stories were multiplied at a conference recently held (December 1968) on "Ecological Aspects of International Development," a brief report of which has been published.[11] A particularly glaring example of failure to look ahead to the probable consequences of a technological change is the sequence of events following completion of the Aswan High Dam in Egypt. We need say nothing of the consequences of shifting the agricultural pattern of an ancient land from dependence upon annual flooding of the Nile to irrigation. Less generally known is the catastrophic effect upon the sardine fisheries of the eastern Mediterranean, since these little fishes have evidently depended for their abundance upon the annual charge of rich organic matter entering the sea when the river flooded, organic matter which led to the prolific growth of the marine plant and animal plankton upon which the sardines feed. Another dangerous consequence of the Aswan Dam has been the rapid spread through the population of infection with parasitic blood flukes, or schistosomes, carried by snails as intermediate hosts. As in Rhodesia, following construction of the great dams there, debilitating and often fatal schistosomiasis has spread rapidly. A third inevitable consequence, which seems little regarded by the builders of dams, is the relatively brief life of the reservoir impounding the irrigation waters. When a river carries a heavy burden of silt, as is so generally the case in arid lands, the reservoir fills up with mud in a few years. Lake Mead is already about half full of sediment. The

[11] Harmon Henkin, "Side Effects: Report of a Conference on Ecological Aspects of International Development," *Environment*, 11, No. 1 (1969): 28–35; 48.

Aswan reservoir will have a very limited life in terms of the stretch of human history. What is to happen when the reservoir is filled and no longer contains a large volume of water, either for irrigation or recreation? No effective means of flushing out the reservoirs has been devised. Do we then blow up the dams and later rebuild them? What happens to our agriculture in the meantime, after it has been transformed to depend upon irrigation? Have we any engineers thinking about the long-range effects, the ramifying side effects, of dams?

Of course, the consequences of introducing a big dam into the landscape depend in very great measure upon the nature of the land where the dam is built, as well as the agricultural practices of the people dependent upon the river. A big dam in Arizona is very different in its effects, and particularly in its side effects and its long-range effects, from the big dam in Egypt or on the Mekong River; and the Aswan Dam and the proposed Mekong Dam will also differ enormously in their ecological and agricultural effects, even though both populations below these sites have depended up till now upon annual flooding of the lower valleys for the fertilization, planting, and watering of their fields. The Nile waters an arid land; the lower Mekong, a plain alternately baked in drought and deluged by monsoon rains. Each situation must therefore be studied as a local system that is itself part of a wider system of interacting regions.

This discussion seems necessary to make clear why the environmental scientist of today insists with ever louder voice that a full systems approach must be taken to any aspect of technological assessment. No piecemeal, limited engineering or physical-chemical approach can suffice. The biological, psychological, and sociocultural aspects must be introduced in the analysis as well; and long-range economic aspects cannot be ignored, in the prospect of immediate gain.

Can these perils be avoided? Our ideal is a people adjusted to its land and its resources, daily increasing its richness of life through the applications of science and technology to its needs. These needs, for each nation, are unique and grow out of its past as well as its present. They are shaped by its culture and its ideals as well as its material resources. As in the evolution of an animal species, its various aspects must change in consonance with each other, for too great a change in one respect, too little in another, may lead to extinction. For such development to take place in the future crowded, technological world we need a true systems analysis of needs, possibilities, and consequences. The national agencies I propose for each country are needed to relate the factors of change inherent in scientific discovery and education to the whole system of a people living in its homeland, and to plan intelligently for its future well-being. These agencies must grapple with the accumulation and organization of scientific knowledge; they must combat the narrowness of the individual specialized scientist and engineer; they must plan ways and means of overcoming educational obsolescence. Above all, they must determine the way to avoid the insidious effects of technology based only on immediate interest.

International agencies of like nature must be created to deal with problems of worldwide scope—the mounting accumulation of DDT, the pollution of seas and atmosphere, the threat of radioactive wastes, and, of course, population pressure and the ultimate peril of nuclear war. All these matters, so eloquently and so forcefully set forth by Andrei Sakharov in his notable statement in 1968,[12] must be the subject of common planning and universal regulation.

The planned utopia may be a delusion, but without scientific analysis and foresight applied as fully as possible to the charting of the future, man obviously courts disaster. There is perhaps still sufficient time in which to act!

## *Science Education and the Future of Science*

In just one hundred and fifty years, and mostly in the immediately past fifty years, science and technology have wrought spectacular changes in human life. Man's power has been increased immeasurably, through enlargement of his understanding of those physical and biological factors with which he must cope in building a world wherein controlled industry and agriculture, medicine, transportation, communication and commerce, automation and the computer provide man with new products, new standards of nutrition and health, and new forms of recreation and enjoyment. When, with Kenneth Thimann,[13] we consider "science as an instrument of service," we may indeed be inclined to discount the jeremiads and predict a glorious continued growth of science and vast, exciting new technological developments as we enter the twenty-first century. Yet never have the social problems of man seemed so critical and so vast, never has the aim of the scientist always to seek the truth for its own sake seemed so out of joint with the search of man for values he is

[12] Andrei D. Sakharov, *Progress, Coexistence and Intellectual Freedom* (New York: W. W. Norton & Company, 1968).

[13] Kenneth V. Thimann, "Science as an Instrument of Service," *Science,* 164 (1969): 1013.

often unable to state but which he recognizes lie beyond the ease and the material riches promised him by science. Science has created a new world; man himself remains essentially the same as a hundred thousand years ago, both in genetic endowment and, over much of the earth, in actual realization of his potentialities. Save for his accumulated capital of knowledge, passed on through education, save for the material capital of civilization which he may utilize, and save for the broadened vision of himself and his fellow beings within the universe, he remains in his emotions, his thought processes, and his desires no different from the first men who grew grain and domesticated cattle, no different from the Cro-Magnon cave painters of animals they hunted, perhaps no different even from the first men who had ceased to be apes and who made tools and weapons and controlled the use of fire. The bitter truth is that man has made a world that seems to be propelled by some inevitability of its own toward change into an environment inimical to man's own hopes and dreams. Science is indeed an instrument of service, but only if its course can be understood and its technological applications regulated and constrained in the service of man. In this matter one must agree with that great teacher of the past generation, Alexander Meiklejohn, when he wrote: "Our final responsibility as scholars and teachers is not to the truth. It is to the people who need the truth." [14]

Science, for the individual scientist, is his quest for the truth and his tested method of reaching it by investigation. But science is also a social process, with a history that reveals man's increasing dependence upon science as the instrument to expand his power, to achieve his goals, and espe-

[14] Alexander Meiklejohn, *Political Freedom* (New York: Harper, 1960), p. 128.

cially to mitigate his ills. As social process, it has not only a history, but also a future—one of promise and peril. The greater the power, the greater the dangers erupting from misuse. The more complex the machinery of society, the more readily a single defect may produce a total blackout. We have created a system in which only an understanding of the complete system will serve. Technological assessment, in other words, requires the most complete imaginable analysis and study of the impact of any single change upon the entire physical and social system that constitutes man's world. It requires the prediction and control of every alteration brought about by the introduction of some new technological step. It may even be possible, as some have suggested, to create new systems in which altered cycles and feedbacks replace those of the older natural environment. Nevertheless, that will first require a very careful study and testing of model systems before we shall dare to meddle with our already troubled world. Fortunately for us, the invention of the computer seems to have come just in time to aid us in such huge tasks of systems analysis.

In the editorial already cited, Kenneth Thimann expressed the view that while "in the past we science teachers have stressed the fascination of science, the unity of science, or the power of the scientific method; it is time now to stress the role of science as an instrument of service and as the means of curing mankind's ills—time to stress the hope of the future as well as the achievement of the past." [15] I agree, but it is not quite enough. The growing hostility to science in our times has root in a supposed conflict of values. It is symbolized by the machine, which unlike man, with his emotions, desires, and goals, does impersonally what it is made to do

[15] Thimann, "Science as an Instrument of Service."

and does it so efficiently that man himself is caught in the meshes of a system geared to production rather than a concern with other social values. The computer, the most manlike of machines in its ability to store information and recall it upon demand, to perform lightninglike calculations, and to systematize knowledge, only aggravates the conflict of values. What is the use of voting in a democratic election, many persons ask, when the computer can already say how the election will turn out and can determine the issue in advance? Yet everyone should know that a machine is programmed to do what men want it to do. It may predict, with poor reliability, on the basis of an input of a few factors. Or it can predict, with increasing success, as the system being studied is more fully represented by adequate sampling of all the elements and factors it contains.

There is thus no real conflict between the instrument and the spirit, for the instrument is the instrument of the spirit. Writing of this in 1953, I said:

> The instrument is the extension of the senses of the scientific investigator and the refinement and enlargement of his crude powers of manipulation. It renders precise his variable and subjective observations, and records in full and permanent form what would perish if it depended upon his memory and his notes. It has often been said that the history of modern science can be read in the invention and improvement of scientific instruments. . . .
>
> Leonardo da Vinci, four and a half centuries ago, well knew the significance of new instruments and implements, and turned his giant talents, from painting the fresco of the Last Supper in the chapel of Santa Maria delle Grazie, to invention and engineering. Was art then simply to be a handmaiden of the sciences? Not at all, for what Leonardo sought in the sciences—in his studies of flowers, of the anatomy of the human heart, of the nature of fossils—was what he sought also in his ex-

periments in painting, so often unsuccessful and uncompleted. The quest of Leonardo was the quest of the scientific spirit "to explore and to understand the universe . . . to grasp the forms and laws of nature and life as they revealed themselves to his alert and penetrating eye." Art thus became the effort to express in spirit the full scope of human knowledge and understanding. That effort by Leonardo to fuse science and art we recognize today as having been the supreme achievement of the Renaissance.

The role of the instrument in furthering this synthesis of which Leonardo dreamed has grown beyond even the bounds of his imaginative genius. Like thousands of other scientists, I can sit in my room before a few instruments, and have the world before me to see and to hear. Like so many other scientists, I find a special pleasure and refreshment in that synthesis of mathematical proportions, rhythms and harmonies with the creative imagination of the artist to make what we call music. To how many of us, then, the greatest triumph of the instrument is represented by the enlargement of the spirit it brings. The radioed word that death has so untimely claimed Kathleen Ferrier, whom we have never seen or heard in person, creates in us a bitter sense of personal loss. Yet the instrument conquers time as well as space and recreates anew at our will the full marvel of her voice and her artistry. Through the pickups, wires, condensers, electronic tubes, and metal diaphragms of our instruments we learn more fully the meaning, so pregnant for these uneasy times, of the words of John Donne:

> No man is an Iland, intire of itselfe; every Man is a peece of the Continent, a part of the maine; if a Clod be washed away by the Sea, Europe is the lesse, as well as if a Mannor of thy friends or of thine owne were; any man's death diminishes me, because I am involved in Mankinde; And therefore never send to know for whome the bell tolls; It tolls for thee.[16]

[16] Bentley Glass, "The Instrument and the Spirit," *Science,* 118 (October 23, 1953): 3A.

It is not enough that in the future the scientist should recognize his dependence upon society for the support of his work at the level of whatever increasing costs will be necessary as scientific exploration becomes more and more expensive. Society too will count the cost, and weigh it against the benefit, though not, I hope, in respect to each individual project but on the basis of the overall cost of supporting basic research in the sciences, since one never knows in advance which projects will produce significant discoveries. Even the wealthiest of nations may not have funds to supply every biologist with an electron microscope, every biochemist with an ultracentrifuge and a protein analyzer, every astronomer with a radio or optical telescope of high resolution, every physicist with a bevatron, and all of them with giant computers. We must face reality and acknowledge that some scientific programs promise greater returns for human society than do others; and if funds are limited there must be allocations. How much more carefully must those nations plan which are still facing the need to develop rapidly both industrial capability and agricultural sufficiency, as well as good health standards and an educational base sufficient to support all of these!

It is not enough for the scientist to recognize his dependence upon society for support, as I have said; it is also necessary for him to recognize his place in science the social process, to assess his contribution to the betterment of mankind. To see this he must reach beyond the immediate consequences of any technological application of his discoveries and strive to grasp its most far-reaching and long-lasting effects upon the terrestrial environment and the social system of man as a whole. To do this is neither accustomed nor easy. Fortunately the computer can help. Yet it depends upon the input. As the computer programmers say: GIGO (garbage in, garbage out).

We must consequently contemplate a new order of education in science, one truly appropriate for the challenge of the next century. It must begin with a study of science as social process: the history of scientific and technological alterations of human society and the consideration of the present and future goals of mankind. To bring to birth a new sort of history of science, far less than now a recital of who discovered what just when; and to originate a new sort of philosophy of science, far less than now a logical analysis of scientific method and no more—these are urgent priorities. Combined, they will provide a study of science in the service of man, combined with a study of the relation of science to man's goals and ethical values. The content of particular sciences can be adapted to meet these priorities, if a teacher really wishes it. In this way, content and process can become fully wedded in the teaching of science; the duality of subject matter and method of inquiry can be wholly united.

It is of course not sufficient for the young scientist himself to be trained in this new way. I am still of the opinion I expressed in an earlier address in the following words:

If we are going to develop a civilization broadly and soundly based upon scientific foundation—and we can hardly escape that now—every citizen, every man in the street, must learn what science truly is and what risks and quandaries, as well as what magnificent gifts, the powers that grow out of scientific discovery engender. Surely, this is our primary task. If we fail in this, then within a few years we may expect the worst, for one or another of these great crises will become a cataclysm. Perhaps nuclear devastation. Perhaps famine. Perhaps a spoiled and stinking environment. Perhaps world-wide tyranny. It is not safe for apes to play with atoms. Neither can men who have relinquished their birthright of scientific knowledge expect to rule themselves. For the scientific society to be democratic and to remain democratic, the people themselves must understand the nature of the scientific forces and problems that dominate their lives. For us who are teachers, this is our task and our commitment.[17]

Here we shall find relief from the trivial. Here we may seek and join the timely with the timeless, the socially relevant with the eternally true, the goals of man with his status in the universe, of which he is indeed so small a part. The task will require a rewriting of all textbooks and a reorientation of every attitude and method in lecture and laboratory. Science curriculum studies will need to discard what they have just completed in order to begin afresh. All levels of science instruction must be changed. The task will be costly and long and hard, but the end is not even the advancement of science, though that will accrue. The true end is quite literally the salvation of man.

[17] Bentley Glass, "What Man Can Be," *Educational Record,* 48 (1967): 101–109.

# INDEX

## ABOUT THE AUTHOR

Bentley Glass is an internationally distinguished geneticist. His numerous scientific, professional, and general publications include the books *Genes and the Man, Science and Liberal Education,* and *Science and Ethical Values*. He has been a leading national figure in the development of new biology curricula for secondary schools. Many universities and organizations have honored him, including the National Academy of Sciences, the American Academy of Arts and Sciences, and the American Philosophical Society. He has held the presidency of many scientific and scholarly organizations, including the American Association for the Advancement of Science, the American Society of Human Genetics, Phi Beta Kappa, and the American Association of University Professors.

Human and social problems have concerned him deeply; he has worked, for example, with the American Civil Liberties Union, and with governmental agencies concerned with the genetic effects of atomic radiation. He was born of missionary parents in 1906 in Laichowfu, Shantung, China. He was a member of the faculty of Johns Hopkins University for eighteen years and has been at the Stony Brook campus of the State University of New York since 1965, where he is Distinguished Professor of Biology and Academic Vice President.

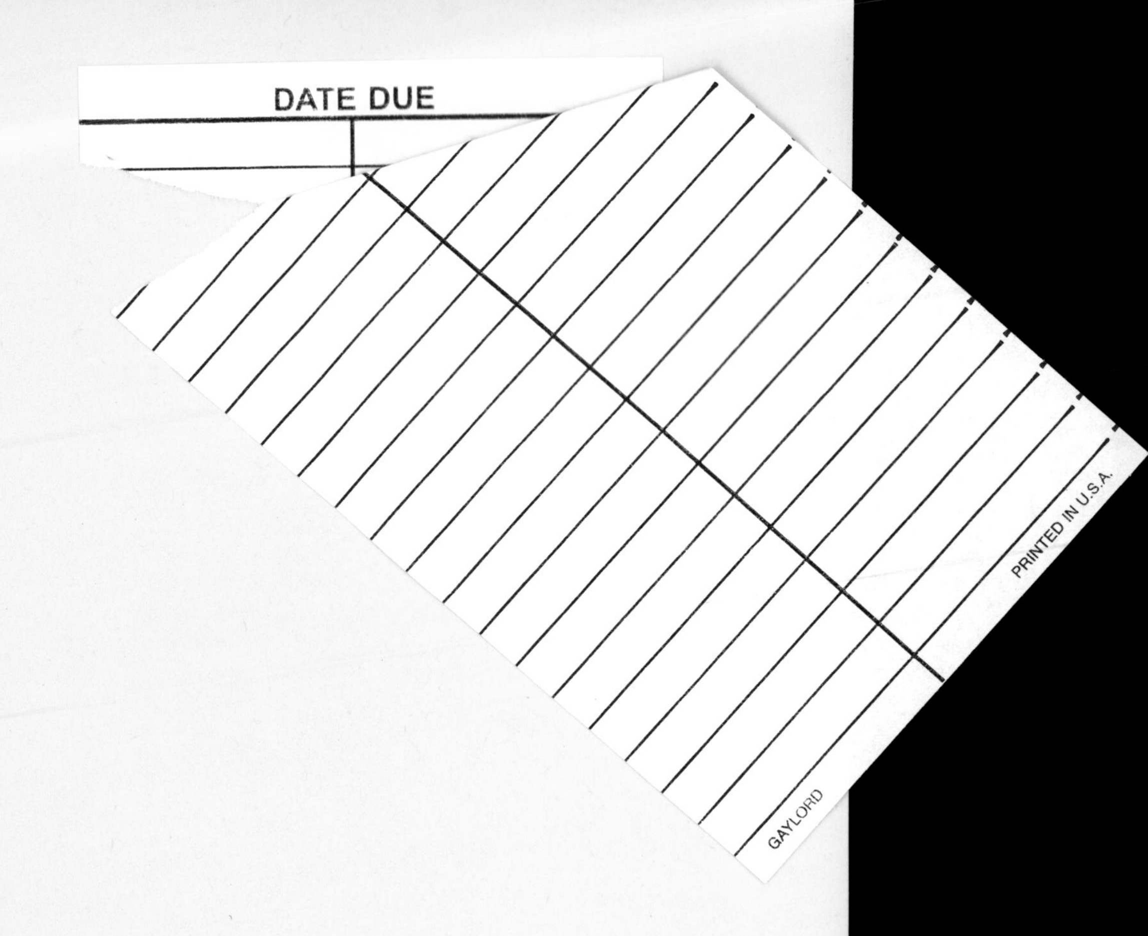
DATE DUE
GAYLORD
PRINTED IN U.S.A.